CORPORATE COMMUNICATION

PETER LANG
New York • Washington, D.C./Baltimore • Bern
Frankfurt • Berlin • Brussels • Vienna • Oxford

Michael B. Goodman
Peter B. Hirsch

CORPORATE
COMMUNICATION

Strategic Adaptation for Global Practice

PETER LANG
New York • Washington, D.C./Baltimore • Bern
Frankfurt • Berlin • Brussels • Vienna • Oxford

Library of Congress Cataloging-in-Publication Data

Goodman, Michael B.
Corporate communication: strategic adaptation for global practice /
Michael B. Goodman, Peter B. Hirsch.
p. cm.
Includes bibliographical references and index.
1. Communication in management. 2. Corporations—Public relations.
3. International cooperation. I. Hirsch, Peter B. II. Title.
HD30.3.G655   658.4'5—dc22   2010008209
ISBN 978-1-4331-0622-4 (hardcover)
ISBN 978-1-4331-0621-7 (paperback)

Bibliographic information published by **Die Deutsche Nationalbibliothek**.
Die Deutsche Nationalbibliothek lists this publication in the "Deutsche
Nationalbibliografie"; detailed bibliographic data is available
on the Internet at http://dnb.d-nb.de/.

Cover design by Clarice Falcao

The paper in this book meets the guidelines for permanence and durability
of the Committee on Production Guidelines for Book Longevity
of the Council of Library Resources.

DEDICATION

To Karen and Deborah, who give meaning to all our endeavors, great and small.

CONTENTS

LIST OF ILLUSTRATIONS

PART 1—Thinking about Corporate Communication

Chapter 1: Adapting to Radical Changes in Business and Media: A Corporate Communication Vision for the Future

Chapter 2: Leadership Capabilities: The Core Competencies for Corporations and Executives

PART 2—Understanding the Forces that Shift the Context of Corporate Communication

Chapter 3: Corporate Communication and Web 2.0

Chapter 4: Strategic Ethical Relationships—Trust and Integrity.

Chapter 5: Corporate Culture's Increased Significance

PART 3—Managing Public Issues: Models for Corporate Communication Practice

Chapter 7: Precedent—The History of Communication in Corporations

PART 4—Strategic Adaptation for Global Practice

Chapter 10: Corporate Communication: The Way Forward

PART 5—Guidelines

PREFACE

The chief communication officer at a Fortune 500 multinational corporation today faces the challenges of a rapidly changing global economy, a revolution in communications channels fueled by the Internet, and a substantially transformed understanding of what a 21st-century corporation stands for.

This book is an attempt to describe these forces and the specific communication challenges that they have thrown at the global corporation. In examining these forces and how they are interwoven, we hope to offer insights and strategies for students of the corporate communication discipline and business leaders to help them deploy effective communication as a strategic business asset in today's global economy. At the same time, we aim to provide concrete and specific recommendations for how to organize and execute effective communication for the contemporary practitioner working in the communication field. The combination of a theoretical framework for understanding how these forces influence corporate communication and practical guidelines for effective communication within this framework will also be of value to students of the communication discipline.

The need for this book arises from the confluence of three forces:

- *Globalization*: a quantitative shift in the globalization of the world economy that has created a qualitative change in how businesses need to communicate;
- *Web 2.0*: a transformation in the adoption, use, and consumption of information technology;
- *Corporate Business Model: The Networked Enterprise:* an evolution in the nature and purpose of the public corporation that is both influenced by and, at the same time, influences the other two forces at work.

The Global Context

Perhaps as an inevitable consequence of the profound transformations that have taken place in the global economy, the role of the corporation, and the advance of information technology, the legitimacy of many key institutions, from global corporations to the governmental and intragovernmental organizations that regulate them, has been called into question. Observers from a wide variety of perspectives question the efficacy of current regulatory frameworks to render the appropriate balance between the rights of the individual investor, a fair return to institutions funded by private capital that stimulate economic growth, and the protection of human and natural resources around the world. Even without the systemic economic events of 2008, these tensions would have continued to throw up important questions about the relationships among different social and economic stakeholders. These questions are having an important impact on how institutions position themselves, how and what they advocate for, and how they view each other. It is this impact, in turn, that is driving shifts in communication strategies. However, when we factor in the twin effects of military conflict and economic turmoil as critical as any since the Great Depression of the 1930s, we have the recipe for a fundamental shift not only in the way organizations communicate, but in the way they behave. In all likelihood, this condition of global anxiety will persist, resulting in a thousand shocks that inhibit "normal operations" permanently.

In this volatile state, corporations and other leading organizations need to make fundamental adjustments in their responsiveness to change, creating, from an organizational perspective, a state of "strategic adaptation." In essence, this means:

- Identifying and managing new global risks, rather than relying on passive compliance models for familiar and established risks;
- Engaging in active dialogue with all public stakeholders in a transparent way to influence rather than control information in order to demonstrate value and assure an uninterrupted license to operate;
- Constructing business operating models that can cope with sudden market, trade, and regulatory shifts to protect against threats and seize opportunities.

This state of "strategic adaptation" applies not only to corporations, but also to governmental and nonprofit organizations. It applies as much to the regulatory bodies themselves as to the organizations they regulate. Organizations all over the world need to create the means of dealing pragmatically with this new reality, and their leaders need to understand the trigger points that signal shifts in the external environment producing new threats, as well as new opportunities.

The conceptual framework for this book is designed to accomplish two principal aims:

- to give corporate and organizational leaders around the world a way to triage evolving issues to help identify points of threat and opportunity;
- to provide them with insights into how the leading practices in corporate communication can help their organizations manage in this state of "strategic adaptation."

This book is intended to provide a useful way of describing what is different for corporations.

It also explores what individuals and organizations can do differently in the face of these changes.

We have built this discussion on three concepts:

- Change causes communication to be even more important than ever.
- The size and scale of global corporations, institutions, and the interconnected business environment present new challenges and opportunities.
- The hyper-connected communication environment has created relationships, challenges, and opportunities that never existed before.

The six parts of this book, therefore, take a strategic communication perspective in its focus on areas critical to the survival of any organization.

Part 1, *Thinking about Corporate Communication*, consists of two chapters: "Adapting to Radical Changes in Business and Media: A Corporate Communication Vision for the Future"; and "Leadership Capabilities: The Core Competencies for Corporations and Executives." Part 2, *Understanding the Forces That Shift the Context of Corporate Communication*, is presented in four chapters: "Corporate Communication and Web 2.0 "Strategic Ethical Relationships: Trust and Integrity," "Corporate Culture's Increased Significance," and "Economic Factors." Part 3, *Managing Public Issues: Models for Corporate Communication Practice*, offers three chapters: "Precedent: The History of Communication in Corporations," "Philosophy: The Engineering of Consent and Process: Strategic and Tactical Models," and "Performance: The Measures That Determine the Success of Communication." Part 4, *Strategic Adaptation for Global Practice*, offers a chapter on meeting the challenges of global business entitled "Corporate Communication: The Way Forward."

Part 5, *Guidelines*, offers more than a dozen tactical discussions on Corporate Communication Strategy and Policy, Crisis Communication, Media Relations, Employee Relations, Global Relations, Corporate Citizenship, Core Competencies for Corporate Communication, Investor Relations and Sustainability, Transparency and Disclosure, Reputation Management, Transaction Communication, Affiliate Relations, and Thought Leadership and Executive Relationship Management. Leading practices in many of these areas are changing rapidly, particularly in investor relations and corporate disclosure. We recommend that readers consult www.sec.gov regularly for the latest changes in U.S. government regulations.

Part 6, *Further Readings and Websites*, provides an extensive list of published sources on how to think about the practice of corporate communication, as well as relevant websites devoted to issues central to corporate communication.

This book defines corporate communication as a strategic management function for making decisions, for creating responses to internal and external audiences, and for developing contingencies to identify and meet new challenges. Radical changes and momentum in the corporate environment challenge leaders to handle all types of communication inside the organization, as well as ever-more-complex relationships with audiences outside it.

Communication technologies and corporate change have created new patterns in communication. For that reason new processes are necessary to use technology for successful long-term impact. This book uses the results of several studies to focus on the "why" of communication.

Success and survival of both the corporation and of individual corporate executives depends on a thorough understanding of the trends in corporate communication, as well as the ability to implement the leading practices that apply to a continuously shifting context.

Understanding and identifying these trends and practices are informed in part from the results of the CCI—Corporate Communication International's studies. CCI's Research includes:

- CCI—Corporate Communication International: Practices and Trends in Corporate Communication 1999–2007; 2009
- CCI—Corporate Communication International: Practices and Trends in Corporate Communication China Benchmark Studies 2006, 2008
- CCI—Corporate Communication International: Practices and Trends in Corporate Communication South Africa Benchmark Studies 2008
- CCI—Corporate Communication International: Practices and Trends in Corporate Communication EU Benchmark Studies 2009–2010

This book also addresses the "how"—the implementation of decisions by offering "Guidelines" for effective and useful practices. More than a dozen guidelines offer leading practices for putting corporate communication strategy into tactical action for use in a working environment or in a university setting for professional orientation and development. It also provides caveats for practice by suggesting situations that might require actions that seem counter-intuitive—especially guidelines for when to close the rulebook and use experience, expertise, and judgment.

The chapters of this book elaborate how social, political, financial, moral, and technical forces are changing the way corporations and individuals must do business to survive beyond the immediate future and thrive well into the 21st century.

ACKNOWLEDGMENTS

The idea for this book came from Frederick D. (Sandy) Sulcer (1927–2004). Sandy was Vice Chairman of DDB Needham Worldwide, having rejoined the agency after 12 years at D'Arcy, Masius, Benton & Bowles. He had begun his career at Needham, Harper & Steers, where he rose to become Director of the New York office and counted "Put a Tiger in Your Tank" among his many creative credits. For many years, Sandy led a lecture series with which we were affiliated entitled "While You Were Looking the Other Way," which sought annually to capture the many subtle changes to the marketing and communication environment over the preceding 12 months that were collectively leading to major cultural shifts. His approach became the inspiration for this book.

We also have many colleagues, friends, and family to thank for their help in bringing this project to fruition. Foremost among them is Tina Genest, associate director of CCI—Corporate Communication International, who tirelessly reviewed every word as it emerged and provided sound critique and much-needed encouragement along the way. We are enormously grateful to the many seasoned professional communicators who took the time to share the fruits of that experience with us, including Jack Bergen, Roger Bolton, Tom Buckmaster, Robin Cohn, John Cox, Stephen Dishart, Emma Gilding, Stu Goldstein, Harvey Greisman, Tom Harrison, Peter Horowitz, Ray Jordan, John Koegel, Roger Leaf, Dick Martin, Catherine Mathis, Mike Maslansky, Mark Monceau, Doug Petkus, Pat Sloan, Jerry Swerling, Reid Walker, and Tom Watson. We are also grateful for the support and encouragement of Jeffrey Peck, Dean of the Weissman School of Arts and Sciences at Baruch College/City University of New York, former Deans Myrna Chase and David Dannenbring, and Jana O'Keefe Bazzoni, Chair of the Department of Communication Studies at Baruch.

We are particularly grateful to the many colleagues, professionals, friends, and students who have so graciously shared their experiences, insights, and comments with us. Our graduate students have been a constant source of ideas, questions, and information, particularly the graduate students in our seminars at Baruch College/City University of New York, Columbia University, Fordham University, New York University, Fairleigh Dickinson University, Aarhus School of Business (Denmark), Hong Kong Polytechnic University, Bangkok University, and the University of Johannesburg. We are particularly grateful to the graduate assistants at Baruch College—Cynthia Chang, Mansura Ghaffar, Darnide Cayo, and Sin Yee Ng—as well as Jenna Gable at IBM and Amy Paulsen.

Thanks to the members of the Advisory Board of the Master's Program in Corporate Communication at Baruch College: Nicholas J. Ashooh, Cynthia Bell, Roger Bolton, Norm Brust, Steve Cody, Linda E. Dunbar, Rachel Lyn Honig, Wendy Kouba, James E. Murphy, Ralph Piscitelli, Jr., Richard S. Roher, Dr. Patricia Scott, Art Stevens, Marc S. Strachan, Loretta Ucelli, Jayne Wallace, Milton M. Weinstock, Kenneth L. Wyse, and Christina Latouf.

Colleagues from the Arthur Page Society have graciously shared their experiences and expertise with us. Thanks to Roger Bolton of Aetna (retired), Dr. James O'Rourke of Notre Dame University, Dr. James Ruben of the University of Virginia, Alan Kelly, Dr. Don Stacks of the University of Miami, Tom Nicholson, Paul Basista, Peter Debreceny of Allstate (retired), Bob DeFillippo of Prudential Financial, Tom Buckmaster of Honeywell, Steve Cody of Peppercom, Kathleen Fitzgerald of KPMG, Jack Bergen of Alcoa, and Tom Martin of ITT. We grieve with Harvey Greisman's family, friends, and colleagues at his sudden death early in 2010.

Participants in the CCI—Corporate Communication International's Conference on Corporate Communication (for the past 11 years in the United States and since 2001 in England) have generously contributed their thoughts and insights. Each has contributed directly and indirectly to this book. Special thanks go to Dr. John Liepzig, Emeritus Professor, University of Alaska Fairbanks; Dr. Krishna Dhir, of the business school at Berry College; Dr. Stacy Connaughton of Purdue University; Dr. Alison Holmes, visiting fellow at Yale University; Dr. Finn Frandsen, Dr. Winni Johansen, and Dr. Frank Pedersen at Aarhus School of Business.

The dedicated professionals we have had the pleasure of working with as part of the PR Coalition have demonstrated the true spirit of volunteerism, especially Jim Murphy of Murphy and Company, who has been a valued mentor and supporter. And thanks to colleagues at Fairleigh Dickinson University—J. Michael Adams, Bernard Dick, Walter Cummins, and Al Schielke. Special thanks to Dr. Nicholas D.J. Baldwin, Dean of Wroxton College, England, for constant conversations over the years about living and working overseas, as well as Trudi Baldwin, director of the Master's programs in communications at Columbia University's School of Continuing Education.

The members, advisors, and sponsors of CCI—Corporate Communication International have been valued sources of support, information, and inspiration, especially Lowell Weiner of Medco Health Solutions; Bob DeFillippo of Prudential Financial; Rich Teplitsky; Ray Jordan of Johnson & Johnson; Martin Hirsch of Roche; John Santoro of Pfizer; James Whaley of Siemens; and Sharon Prince and Wendy Kouba at Wyeth.

Sincere thanks to David Milley, webmaster for CCI—Corporate Communication International for his dedication, knowledge, and insight.

We would also acknowledge the following for granting permission to print: Wendy Kouba for Figures 2.4 and 2.5; Johnson & Johnson for Figure 4.1; Oxford Metrica for Figure 9.1; and for permission to quote from their interviews with us - Catherine Mathis, Dick Martin, Ray Jordan, Reid Walker, Tom Buckmaster, Jack Bergen, Tim Hedley, and Harvey Greisman.

Thanks to Nanci Healy, editor of the *Journal of Business Strategy*; to Wim Elving, editor of *Corporate Communication: An International Journal*; and to Dr. Sandra Oliver, former editor of *CCIJ*.

Thanks also to Mary Savigar at Peter Lang for having asked us if we had a book on corporate communication that we would like to write, and for believing in this project.

Thanks to Harold A. Goodman and Ann Kahn.

Finally, thanks to our wives Deborah Hirsch and Karen Goodman, our best critics, editors, and friends; and to our sons and daughters J. David Goodman and Craig M. Cook, Tim and Becky Hirsch, who may take for granted their roles in our world.

Michael B. Goodman
Peter B. Hirsch
New York City, 2009

PART ONE

Thinking about Corporate Communication

In order to understand how radical changes in business and the media have influenced the practice of corporate communication, we need to look first at those changes themselves. In particular, we need to reflect on the relationship between the behavior of multinational corporations and the political world order following World War II. The period from 1945 to 1975 was not only a period of unprecedented economic dominance of the world marketplace by the United States but was also the apogee of the American way of doing business. American management practices embodied in giant world-spanning enterprises such as DuPont, Ford, ExxonMobil, General Electric, and Procter & Gamble became the gold standard for multinational corporate practices.

Not surprisingly, the American approach to the practice of the corporate communication discipline also became the subject of study and emulation throughout the world. It was this intertwining of American commercial and economic dominance with its geopolitical hegemony that caused Charles Wilson, CEO of General Motors, to say, when nominated in 1953 to be President Dwight D. Eisenhower's Secretary of Defense, "For years I've thought that what was good for the country was good for General Motors and vice versa." It is important to understand the embedded strength of this mindset in order to gain insight into why it persisted long after the geopolitical and economic landscape started to be transformed in the 1970s. It is also highly instructive to retrace our steps through the rise of the Japanese economy in the 1980s and the emergence of a truly "unipolar" world brought on by the ideological and geopolitical collapse of the Communist Bloc in the 1990s, all of which played a key role in the evolution of the multinational corporation.

We also need to describe the nature of the far-reaching technology changes that have transformed communication media in the last two decades, starting with Tim Berners-Lee and his 1989 proposal to CERN that led to the World Wide Web. We need to look at the scale, reach, transparency, persistency, and connectedness of a series of technological innovations spawned by the Internet in order to understand the new opportunities made available to the communication professional and the new reputational liabilities to which it exposes the modern corporation. We illustrate these changes with examples from the world of one leading institution, IBM Corporation, showing how it has adapted its corporate communication and other institutional behaviors through the use of these new tools.

Also, in this part, we examine the ways in which changes in the global economic system have combined with technology change to alter the role and structure of the multi-national corporation. We show how the emergence of truly global supply chains, enabled by globe-spanning technologies but abetted by beneficial geopolitical developments, set in motion a profound re-examination of the fundamental management questions: In bringing my products to market, what do I need to own? What can I buy from others? What should I control? Where do new ideas come from? How do I finance my business? In discussing the new answers to these questions that modern corporations found, we can show how the changed power relationships between companies and their stakeholders, indeed the emergence of entirely new stakeholders, set the stage for a significant shift in the theory and practice of global corporate communication.

This part explores the evolving discipline of corporate communication and how corporations adapt to influences bringing about change. We examine the impact these forces have on how corporations and individual executives practice corporate communication and how corporations adapt their management processes and structures to adapt to the radical changes. This part identifies and analyzes the core competencies and leadership capabilities that have emerged as a consequence of the global shifts in the business and media environments.

CHAPTER ONE

Adapting to Radical Changes in Business and Media

A Corporate Communication Vision for the Future

To understand how radical changes in business and the media have influenced the practice of corporate communication, we need to look at those changes themselves. We need to reflect on the relationship between the behavior of multinational corporations and the political world order following World War II ("Globalization," below). We also need to describe the nature of the far-reaching technology changes that have transformed communication media in the last decade ("Web 2.0"). And we need to examine the ways in which changes in the global economic system have combined with technology change to alter the role and structure of the multinational corporation ("Corporate Business Model: The Networked Enterprise"). Constructing this framework will enable us to take up the role of corporate communication and how it should be defined in order to address how the forces we describe have shifted the context of corporate communication.

Adapting to Radical Changes in Business and Media

The transformation of the world economy over the past decade, however startling, is in reality an outgrowth of many forces that have been developing over a much longer period of time. To many observers, the emergence of powerful global companies in what were until recently called Third World countries has been a shocking change of very recent vintage. Viewed from this perspective, the dominance of North American and Western European (and, later, Japanese) companies in the world economy was deemed to be a feature of an enduring and fundamental world order. In reality, the destruction of most of the world's productive capacity outside the United States during World War II and the subsequent isolation of the economies of the Communist Bloc between 1945 and the early 1990s created the relatively brief but vast disparity between "developed," "developing," and "undeveloped" economies that characterized the golden age of American economic hegemony.

Globalization

Even after the economic downturn of 2008–2009, the U.S. economy continues to be larger than the economies of the emerging nations of China, India, and Brazil. Nevertheless, it was this historical economic disparity that formed the worldview of multiple recent generations. In this worldview, the global economy had a clearly identifiable and simple framework: raw materials in the form of energy, metals, food, and other agricultural commodities, such as timber and cotton, were sourced from poor or undeveloped countries, transformed into higher-value products through the superior scientific and manufacturing prowess of the developed nations, and consumed by the populations of these same countries. To the extent that productive capacity was located in the developing world, this capacity was largely controlled by or served the needs of global corporations headquartered in Europe and North America.

The generations born after 1945 also grew to expect that the global institutions and frameworks that managed the world economy would also be controlled by the same nations with dominant productive capacity. Thus the United Nations, GATT (now the WTO), the International Monetary Fund, the World Bank, and Bretton Woods, in alliance with North American and European corporations, were the rightful and autocratic stewards of the world economy. The excesses of this regime, such as the corruption of political institutions in "source" nations by the United States to serve the aims of American corporations, were first assailed by Western critics decades ago. However, the fundamental rightness of this global structure was never seriously questioned by most of the consumers of the dominant powers, who benefited from decades of rising prosperity, plummeting mortality rates, and increasing time off from labor for leisure and entertainment. After the end of the Cold War in the 1980s, these citizens could add to their blessings an unprecedented era of geopolitical stability that lasted until the eruption of the terrorism associated with Islamic fundamentalism in 2001.

This highly selective and artificial survey of the economic and geopolitical history post-1945 obviously leaves out significant events and transformations of this era. The post-colonial emergence of independent nations in Africa and Asia, the wars for Israel, the Communist takeover in China, Vietnam, the agricultural revolution of the 1960s, and the oil crisis of the 1970s are only the most obvious events that do not fit into the portrait of uninterrupted "Western" triumph and dominance that we have painted. Nor have we taken into account the assassinations of presidents and CEOs, the turbulence of 1968, or the Iranian hostage crisis. What we are trying to suggest is that, these cataclysmic events notwithstanding, the fundamental economic organization of the world was profoundly tilted in favor of a few privileged nations.

Understanding this imbalance, as we might describe it from the perspective of 2009, is consequently vital to understanding how communication pathways were shaped up to the present. In painting this picture, we are simply making the argument that the global corporation, almost exclusively North American, European, and Japanese, shaped its communication to meet the needs of this economic framework. In this framework, the style, language, content, and form of corporate communication was designed to nourish relationships with and serve the needs of Western consumers, workers, investors, regulators, and legislators, local communities, and global institutions controlled and influenced by these stakeholders. The emergence of powerful non-governmental organizations (NGOs) in the health and environmental spaces during the 1970s and 1980s, while providing new challenges to corporate communicators, did nothing to fundamentally change the focus on these core stakeholder groups.

We have exaggerated certain aspects of this historical portrait precisely because it enables us to make a starker contrast with the economic environment as we observe it today. Whereas in 1945, 75 percent of the world's productive capacity emanated from the United States, today that number is down to 13 percent and falling year by year. Sales figures for U.S. companies show an analogous pattern, with foreign sales growing steadily as a percentage of the total over the past few decades. This trend has even accelerated in recent years. According to Standard and Poors, foreign sales for U.S. companies in 2006 represented 44.2 percent of the total, up 37 percent since 2001. In 1970, General Motors employed 500,000 workers, of whom 350,000 went to work every day in 25 U.S. and Canadian auto plants. Even prior to that manufacturer's travails in 2008, 60 percent of those same workers toiled in 50 plants from Shenchen to Zaragosa. China is already the largest mobile phone market in the world, having surpassed combined sales for the United States and Europe in 2002. Sovereign wealth funds, the largest class of non-U.S. investors, accounted for $3.22 trillion in assets invested (Sovereign Wealth Funds 2009 estimate according to Prequin, research consultants). Within the past decade, millions of Indians and Chinese have entered the middle class. According to the McKinsey Institute, that growth is expected to continue, with India's middle class growing from 50 million to 583 million in the next 20 years. During the same timeframe, according to the institute, China's middle class will grow from 43 percent to 76 percent of the population.

The story of this shift has been recounted elsewhere more eloquently than we could possibly attempt. Such widely acclaimed works as Tom Friedman's *The World Is Flat* and Joseph Stiglitz's *Making Globalization Work* are excellent sources for further detail on the great transformation of the world economy. We provide just enough detail to underline the vastly different communication challenges that corporations will face. We go into more detail on these challenges in ensuing chapters, but the questions they raise merit a mention here: What does it mean for corporate communication when a majority of its employees have limited personal access to the Internet? What are the ramifications of having a predominantly secular or Judeo-Christian customer base in contrast to one that is overwhelmingly Islamic or Hindu? What are the public affairs consequences of moving from doing business in countries with a predominantly democratic polity to countries that are theocratic or one-party states? How should equitable career tracks be established in countries in which the very government itself is constructed on the basis of an ethnic quota system?

We are well aware that today's global corporations are not facing these issues for the first time. Some of them have operated in culturally and politically distinct markets since the mid-19th century. Our argument is simply that it makes a profound difference to the corporate brand when these diverse publics are not just marginal but core to the future of the enterprise. As we will show in later chapters, global corporations will need to reinvent how they communicate in order to account for this permanent shift in stakeholder dynamics.

Web 2.0

The second force, the emergence of broadband Internet and the functionalities that are often collectively referred to as Web 2.0, plays a curious double role in the transformation of corporate communication. Because it is a cause, its existence requires a new approach to communication and a tool to respond to the very challenges to the corporation that it represents. The speed with which this medium has grown and the scale it has now achieved exceed all conceivable superlatives. In 2008, according to Wikipedia, Google engineers made it known that Google Search had identified one trillion unique URLs. In the same article about the World Wide Web, Wikipedia cites DomainTools as the source for the statistic that measures the number of domain names as 100.1 million, 74 percent of which are commercial in nature.

However, it is not the raw scale of the medium that is significant as much as the number of human beings who are connected by it. The "World Internet Usage Statistics" site run by the Miniwatt Marketing Group, which aggregates information from a variety of public sources, shows that in 2008, 1.4 billion individuals around the world had Internet access. These individuals accounted for 74 percent of the population of North America and 48 percent of the population of Europe. Don Tapscott and Anthony Williams in *Wikinomics* describe the web as the world's largest coffee house in which there are 50 million bloggers sharing their opinions. Between these bloggers and other individuals posting to blogs, there are, say Williams and Tapscott, 1.5 million posts a day. Furthermore, these interlocutors are not restricted to simple text. In an increasingly large part of the world, broadband access makes it possible to view and post still images, moving images, audio, even software applications at zero marginal cost. With these tools, both champions and critics of corporations and other organizations can share views both laudatory and loathsome with an audience that can number in the millions.

The power that Web 2.0 puts into the hands of a company's stakeholders is not just restricted to this newly available ability to post information to a large audience. There are a variety of other features of this new medium that make it exponentially harder for companies to manage both the information about themselves and their relationships with stakeholders. The most important of these is the fact that, as governments—as well as corporations and individuals—go online, they are loading statutory public data to their organizational websites. Information about business transactions, labor practices, environmental compliance, and intellectual property filings that was once available only through lengthy and cost-prohibitive visits to governmental records offices is now instantly accessible to anyone with an Internet connection. Not only are individual corporate records now available, but in many cases software applications have been written that enable consumers, investors, and critics to instantly compare the records for one company against another or against standards espoused by NGOs.

Almost as critical is the way that Web 2.0 actually functions. Although the term "social media" is currently used to describe the mechanisms by which individuals engage in private or personal interactions online, it is actually a very accurate description of what the entire web now enables: social interactions around specific themes or interests. Through search, tagging, RSS feeds, "friending" sites such as Facebook and MySpace, "best of" lists and recommendations, discrete communities of individuals with often-strong opinions about corporate activities are being created every day. Not only are they created, but the automated nature of the applications involved means that these communities are sustained with very little effort on the part of the membership. It only requires a tiny number of obsessive data miners, linkers, and posters to keep a community of millions alive and engaged.

This instantaneous and persistent engagement creates a speed of response that can be breathtaking. The YouTube video illustrating how to pick a Kryptonite bicycle lock with a ball point pen was seen by millions within days of being posted. When Facebook altered its settings to permit friends to find out what their friends were buying online, within a week 50,000 Facebook users became members of a protest group.[1] A week later, *The New York Times* (November 29, 2007) reported that Facebook CEO Mark Zuckerberg had reversed the policy change. Within hours of airing a news story featuring a 1960s document related to President George W. Bush's service in the Alabama Air National Guard, CBS was being challenged on the document's authenticity by experts in 1960s typewriters, who proved that the typeface in the document couldn't have stemmed from a typewriter then manufactured. For corporations used to being the true experts in their own fields, the reach created for external experts by the Internet is yet another complicating factor in managing their reputations.

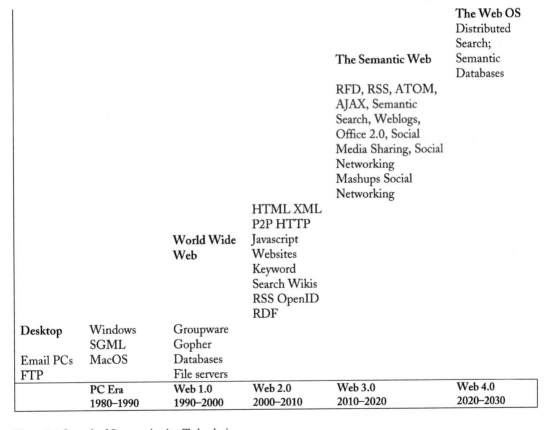

Desktop		World Wide Web		The Semantic Web	The Web OS
					Distributed Search; Semantic Databases
				RFD, RSS, ATOM, AJAX, Semantic Search, Weblogs, Office 2.0, Social Media Sharing, Social Networking Mashups Social Networking	
			HTML XML P2P HTTP Javascript Websites Keyword Search Wikis RSS OpenID RDF		
Desktop	Windows SGML	Groupware Gopher			
Email PCs	MacOS	Databases			
FTP		File servers			
	PC Era 1980–1990	Web 1.0 1990–2000	Web 2.0 2000–2010	Web 3.0 2010–2020	Web 4.0 2020–2030

Figure 1.1 Growth of Communication Technologies

The Case of IBM

IBM is a good example of a corporation that has made successful use of technology to leverage its corporate culture and reputation into a model for sustainable enterprise. It was founded in 1889 as a company that made tabulating machines. Many companies did so as well, but IBM's long-term vision and the reputation it earned were the foundation for its sustainability. It also has a strong and identifiable corporate culture that contributes to its triple bottom line—social, environmental, and financial performance. It places a strong emphasis on its human capital, its Research and Development and Innovation, its leadership and strategic vision, and its emphasis on the environment.

Examples of its strategic vision are reflected in its leveraging of its more than 300,000 employees worldwide through communication technologies:

- BluePedia—searchable IBM wiki used as collective expertise system
- Fringe—social profiles used to locate expertise
- Beehive—like Facebook; social networking to share ideas, expertise, and to test theories

Further, IBM uses communication technologies as the fundamental approach to leadership and strategy. It created "Jams," for example, a computer-mediated brainstorming session on a particular theme, like a jazz jam-session. In 2003, CEO Samuel J. Palmisano initiated a 72-hour "Jam" on the impact of IBM values on client service. It was open to 319,000 IBM-ers. The feedback was direct and revealed that many employees felt that the company's bureaucracy and related dysfunction prevented optimal client service. The Jam caused IBM executives to review the company's corporate values and then make a fundamental shift in its culture. Executives processed feedback and adopted cultural changes that resulted in greater employee empowerment, the institution of participative management, and the creation of executive and corporate cultural self-assessment processes.

One other profound change was the application of this internal capability to its customer relationship. The result was PartnerWorld—social networking for supply chain/partners to collaborate on business opportunities, green initiatives, and market strategy.

IBM's culture of innovation and R&D has resulted in the corporation's emergence as a leading provider of systems to manage and coordinate all the resources, information, and functions of a business. It is a company that uses the technologies and management services it offers its clients and partners. It is ideally positioned for Web 3.0 as a leading provider of "semantic" web applications.

Many companies use Web 2.0 technologies extensively, some even exclusively. However, few multinational corporations have so harnessed human capital or embraced the cultural changes necessary to do so as has IBM.[2]

Reputation + innovation + sustainability = successful strategy for strategic adaptation.

On the other side of the coin, Web 2.0 would appear to give corporations and other organizations large and small access to an unparalleled set of new tools for communicating with stakeholders: corporate text blogs, podcasts and webcasts, RSS feeds for investors, paid search, meta tagging, corporate Facebook profiles, not to mention mobile communication, text messaging, and microblogging. Just as corporate critics are using search engines and blogs to find negative information and link with other critics, corporations now have blog search tools such as Technorati and Ice Rocket to hone in on emerging issues and interact quickly with critical voices. With a few exceptions, however, corporations have not yet found firm footing in this new space and are struggling with how to handle the tsunami of interactions presented by Web 2.0.

The elements of Web 3.0 and Web 4.0 have been available for some time. They form the strategic planning for many multinational corporations.

Corporate Business Model: The Networked Enterprise

The third of the forces transforming communication has been closely interwoven with the transformation of the world economy. It both springs from and to some degree gave birth to the globalization movement and relates to a significant shift in the fundamental function and structure of the corporation. As we will demonstrate, this shift has profoundly changed the relationship between a corporation and its stakeholders to such a degree that a wholesale re-examination and re-orientation of a company's communication with these stakeholders inevitably flows from it.

The joint stock corporation, which was originally established to fund voyages of exploration and helped establish the great trading empires, has come a long way from the day when Samuel Johnson could declare that "the purpose of a corporation is to help a gentleman decide which of his debts he is going to pay." Up through the 20th century, it served this funding and risk management role admirably. With the advent of industrial-scale manufacturing and the development of primary industries in steelmaking, chemicals, and energy extraction, the value and purpose of the corporate structure underwent a significant shift. While raising and managing capital remained important, the corporation took on the additional role of managing the increasingly vast physical plant required for the production of the steel to build railroads and the crackers, refineries, and converters employed in chemicals.

The pioneers of industrial management—the Carnegies, the Fricks, and the Siemens—developed business processes and management disciplines, some of which survive to this day. At the same time, the relatively high cost of transportation created a competitive advantage for companies located close to their markets, energy, and raw materials. Out of these market realities evolved the giant industrial agglomerations of Pittsburgh, Bethlehem, Lancashire, and the Ruhr. As the scale and the pace of industrial production intensified even further, the risks inherent in any interruption to the supply or cost of energy and raw materials grew exponentially.

By the middle of the 20th century, it was not unusual for a major corporation to own and manage every single element in its supply chain, from raw materials to shipping of the final product. The Ford Motor Company's Rouge River plant provides a compelling example of this type of integration. Employing more than 100,000 people in the 1930s, the plant had its own dock on the Rouge River, 100 miles of internal railroad lines, and 16 million square feet of manufacturing space. In 1922, Henry Ford acquired his own coal mine in Kentucky. The wood for the 1928 Model A came from Ford's own forests, and its ownership of mines enabled the company by the 1930s to go from iron ore extraction to a completed vehicle in a freight car in a then-unthinkable three days and nine hours. In one form or another, this model of vertical integration survived into the 1980s. Even as transportation costs fell and light manufacturing grew alongside primary production, the security needs of the Cold War military-industrial complex helped dictate the survival of the vertical model. It also behooves us to remember that in the absence of networked computing power, the financial and administrative functions of a giant corporation could not easily be served without physically centralized document production and retention.

As might be expected, the communication departments of such vertically integrated entities were designed to serve the needs of company towns and their employed populations. In an era in which lifetime employment with a single employer was the norm rather than the exception, and in which business relationships lasted across generations rather than across quarters, the style of corporate communication, even when deployed in acrimonious labor or business disputes, tended to reflect deep and intimate knowledge of the corporation about its stakeholders, and vice versa. The style of communication was also suited to an era when nation-based business structures, Pfizer India Ltd. or Siemens USA, were dominant, and very few messages had to flow across borders.

The contrast with today's business environment could scarcely be more striking. Container shipping has reduced the cost of transportation for both raw materials and finished products. Improvements in information technology have eliminated the relevance of geographical location in the administrative and financial function. The automation of almost every aspect of industrial production has reduced the need for craftsmen trained in a company's own proprietary methodologies. Improvements in the functioning of the capital markets have also contributed to a reduction in the cost of capital, making it easier for companies to invest in building "green field" facilities from scratch. Improvements in technology, combined with reduced transportation costs, even made one of the main costs of manufacturing—energy—fall in an almost straight line from the 1950s through the beginning of the new millennium.

In the context of these reductions in the cost of doing business, it became glaringly obvious that labor costs were accounting for an increasingly sizable percentage of overall corporate expenditures. Faced with increasing competition from the first wave of Asian tigers—Japan, Korea, and Taiwan—American and even European companies began what might be called the great unraveling. First to go were products that were relatively easy to assemble, where shipping costs for raw materials and finished products were already low: textiles, clothing, furniture, and footwear. The explosion in consumer electronics of the 1980s fueled originally by the Japanese decision to enter that market in the 1950s moved the production of televisions, radios, and other audio equipment offshore to Mexico, Taiwan, Korea, and Japan, so that by the year 2000 not a single television set was manufactured anywhere in the United States.

Coterminous with the relocation of manufacturing jobs outside the United States and Europe was the academic re-examination of the business strategy that had supported the idea of vertical integration as well as the industrial conglomerate common in the 1960s and 1970s. First articulated in 1990 by G. Hamel and C. K. Pralahad in their *Harvard Business Review* article "The Core Competence of the Corporation," the concept of core competencies (those activities that give a company its competitive advantage) accelerated the pace of what became known as outsourcing. In the course of little more than a decade, so-called non-core activities were outsourced to dozens of different kinds of specialists. Companies whose own employees once managed transportation, security, waste disposal, cleaning, corporate travel, and a host of other services outsourced these functions as quickly as they could. As information technology became standardized on a smaller number of key platforms, some companies began outsourcing their IT operations. Others looked at strategic but labor-intensive activities such as warehousing, finance, or human resources and decided that these, too, could be handled by outsiders.

To some extent, it became inevitable that a few companies, abetted by ever-cheaper technology, would then outsource and "off-shore" what would once have been regarded as a sacrosanct core activity: customer service. While there have been a few widely publicized examples of companies bringing customer service back on shore, the trend toward having North American or European customer calls being answered by consultants in India, Ireland, or the Philippines appears unstoppable.

Finally, as the 21st century dawned, large global companies began to explore the once unthinkable: outsourcing research and new product development to third parties both inside and outside their national borders. This development was not entirely a response to cost pressures and the markets' insatiable demand for greater profitability. It was also a response to a perceived slowing of scientific innovation after a decade of remarkable scientific achievement in every field from health care to oil exploration. Even Procter & Gamble, a pillar of the traditional U.S. economy, recently announced as a declared goal that by 2011, 50 percent of its new products will have been inspired by innovations sourced from outside the company's four walls. Since the beginning of the new millennium, we have truly entered the realm of the virtual, networked enterprise.

The impact of this hyper-extended global value chain on the world of corporate communication is hard to overestimate. Where once companies were attached by physical proximity to the communities in which they manufactured their products, today they may only be dimly connected to the place in which 80 percent of the value of those products is manufactured. Where once a unitary corporate culture—the HP principles, Dana Style, the Johnson & Johnson credo—held sway in the enterprise, today the leadership of large organizations will have only limited control over or understanding of the human beings all over the globe who assemble, ship, repair, and service the branded products that represent their value in the marketplace. This will require a complete shift in communication strategy, policy, and practice.

The confluence of these three forces—Globalization, Web 2.0, Corporate Business Model:

The Business Challenges for Corporate Communication

The challenges facing multinational corporations have only deepened. It is as if the world economy, like a Fourth of July rocket, burned brightly and then abruptly fell to earth on September 15, 2008, taking Wall Street with it and plunging the world into a deep recession. Jim Collins's *Good to Great* (2001) cites company examples that embodied that transformation, among them Circuit City and

Fannie Mae. Both failed in 2008. And Freddie Mac—which was #28 on the list of 100 Best Companies to Work For, behind #6 IBM and #21 J&J and ahead of #30 3M and #36 American Express—was taken over by the U.S. government along with Fannie Mae in 2008.

Richard Foster, a McKinsey director from 1982 to 2004, described the boom-and-bust business cycles of creation and destruction in his 2001 book *Creative Destruction*. In *The New Paradigm for Financial Markets* (2008), George Soros calls these boom-and-bust forces "reflexivity."

> Reflexivity can be interpreted as a circularity, or two-way feedback loop, between the participants' views and the actual state of affairs. People base their decisions not on the actual situation that confronts them but on their perception or interpretation of that situation. Their decisions make an impact on the situation (the manipulative function), and changes in the situation are liable to change their perceptions (the cognitive function).[3]

Reflexivity was a term also used by Joseph A. Schumpeter in *Capitalism, Socialism and Democracy* (1942).[4]

The financial crisis of 2008 forced a transformation in fundamental business models, beginning with the momentary disappearance of an entire industry sector—investment banking. Manufacturing and the auto industry teetered on the edge as Congress debated its fate, and the media business faced an uncertain future as The Tribune Company filed for bankruptcy on Monday, December 8, 2008.

If that were not challenge enough, corporations and organizations have been facing other obstacles for several years. Skepticism and distrust of corporations (and all large institutions, for that matter: government, the church, and organized baseball) continues to be very high among the general public and the media. Media democratization as a result of Web 2.0 will soon be even more transparent with the new Securities and Exchange Commission requirement for XBRL (eXtensible Business Reporting Language) in financial reporting, which will allow electronic searches of data similar to that of text. The economic disparity between the haves and the have-nots will create an unstable civil society.

Digital technologies and the way people work have altered the building of relationships within corporations and with their external constituents. Talent no longer owes loyalty to any corporation. This is particularly evident in the Millennial Generation, according to Ron Alsop's *The Trophy Kids Grow Up* (2008). However, the global recession of 2008–2009 promises to blunt their apparent sense of entitlement.

In an environment where people trust corporations less, the power of NGOs to influence change has increased dramatically. Finally, there are new issues created by global growth, such as scarce resources, from food to potable water. And global growth has elevated expectations in China, Brazil, India, Russia, and the "Next 11"—Bangladesh, Egypt, Indonesia, Iran, Mexico, Nigeria, Pakistan, The Philippines, South Korea, Turkey, and Vietnam.

Multinational Corporations are buffeted by the forces of CSR (Corporate Social Responsibility) and ESG (Environment Society Governance) (see Figure 1.2). The uproar over CEO and executive pay and use of corporate jets exemplifies the discontent.

The financial turmoil has rekindled the issues of politics and regulation as a result of more government "ownership" of banks, insurance companies, and automobile manufacturers, which will necessitate more collaboration among government, corporations, and labor. Cultural and social acceptance of ideas and behavior such as diversity in the workplace, as well as an acceptance, with the rise of China, that democracy and commerce don't necessarily require one another for success are rewriting our understanding of the relationship between commerce and civil society.

Success also depends on new thinking for sustainable enterprises fueled by business innovation, research and development, and the use of technology—local is both local and global. And access to food, clean water, and breathable air are enormous challenges and opportunities. There is an urgency to rebuild trust in business by management and staff through leadership initiatives and strategic action.

The demand for quality—"Better, Faster, Cheaper" products and services—remains ubiquitous. A reinvented media landscape, combining greater ease in the creation of information with fewer traditional outlets, poses more challenges and opportunities.

And the initial interpretations of globalization (for example, to catch salmon in Alaska, ship it to China for filleting, and then ship it to stores and restaurants in New York) may no longer be sustainable. In the spring of 2008, when fuel prices spiked above $4.00 a gallon in the United States, California strawberry farmers bought land in New England to short-circuit transportation costs to their East Coast markets.

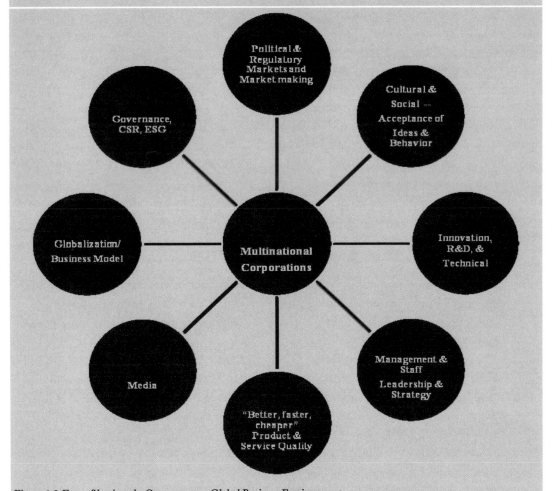

Figure 1.2 Forces Shaping the Contemporary Global Business Environment

The Networked Enterprise—has produced an environment in which most corporate communication practice has become ineffective and outmoded, if not unintentionally destructive. What
is needed is a clear and comprehensive reinvention of the discipline to ensure that a company
can have relationships with its stakeholders, old and new, that will create loyal and enthusiastic
consumers, shareholders, communities, and employees, while persuading NGOs and governmental actors that it has earned and deserves an unquestioned license to operate. This book's later
chapters are our blueprint for this transformation.

A Corporate Communication Vision for the Future

These three words—"how," "authenticity," "adaptation"—inform the vision of a sustainable,
global practice of Corporate Communication.

In 2007 a book with the title *How* appeared, with the subtitle *Why HOW We Do Anything
Means Everything…in Business (and in Life)*. Dov Seidman, CEO of LRN, wrote the book.
Thomas L. Friedman, author of *The World Is Flat* (2005) and *Hot, Flat, and Crowded* (2008),
praised the book when it was published, as well as in his October 15, 2008, *New York Times* op-
ed column, "Why How Matters." Friedman praised it again, quoting Seidman, who argues that
in our hyperconnected and transparent world, how you do things matters more than ever,
because so many more people can now see how you do things, be affected by how you do things
and tell others how you do things on the Internet anytime, for no cost and without restraint. "In
a connected world," according to Seidman, "countries, governments and companies also have
character, and their character—how they do what they do, how they keep promises, how they
make decisions, how things really happen inside, how they connect and collaborate, how they
engender trust, how they relate to their customers, to the environment and to the communities
in which they operate—is now their fate."

The Arthur W. Page Society report, *The Authentic Enterprise*, is must reading for corporate
communication undergraduate and graduate students and practitioners around the world. It presents this path to success:

So, to be an effective communication function in the authentic enterprise:

1. **We must not only position our companies, but also help define them.** While expertise and authenticity are essential, communicators' counsel to the corporation must
   now encompass its fundamental business model, brand, culture, policies and, most
   importantly, values.
2. **We must not only develop channels for messaging but also networks of relationships.** In a business ecosystem of proliferating constituencies, communicators must
   lead the development of social networks and the tools and skills of relationship building and collaborative influence—both to seize new opportunities and to respond to
   new threats.
3. **We must shift from changing perceptions to changing realities.** In a world of radical transparency, 21st century communication functions must lead in shaping
   behavior—inside and out—to make the company's values a reality.[5]

Multinational and publicly traded companies are in the vanguard of the extraordinary transformation of global business. The changes corporate communicators face can be overwhelming, or they can sharpen our focus and resolve. And the capability for "adaptation" is essential to survival. It is important to think long-term, as well as survive short-term.

What is Corporate Communication? What are the challenges corporations face in the 21st century? What is the contemporary business environment? What drives value for contemporary businesses? What are the skills needed to meet these challenges? What are the strategic functional areas of corporate communication identified by the CCI studies? And what are some successful behaviors that speak to how to engage with people inside and outside of your company to meet the sometimes overwhelming challenges of our time?

At the end of the 19th century and early in the 20th, John D. Rockefeller, the founder of what is now ExxonMobil, had a problem relating to the people of New York. His advisor Ivy Lee counseled Rockefeller to give nickels to the children he encountered on his walk to work. These were indeed inauspicious beginnings for our profession. Arthur W. Page, by mid-century, was counseling his colleagues at AT&T, and the society that bears his name has codified his advice in "The Page Principles":

- Tell the truth.
- Prove it with action.
- Listen to the customer.
- Manage for tomorrow.
- Conduct public relations as if the entire company depends on it.
- Realize a company's true character is expressed by its people.
- Remain calm, patient, and good-humored.

Then people began to use the term "Corporate Communication," for which there are now several definitions (see the sidebar below). For this book we will use this one:

> Corporate communication is the term used to describe a variety of strategic management functions. Depending on the organization, corporate communication includes: public relations; crisis and emergency communication; corporate citizenship; reputation management; community relations; media relations; investor relations; employee relations; government relations; marketing communication; management communication; corporate branding and image building; advertising.[6]

Defining Corporate Communication

Paul Argenti, in *Corporate Communication* (4th ed., 2007), has clearly led the discussion of the development of the discipline of corporate communication as a strategic management function.

> …over half of the heads of corporate communication departments oversee communication functions that include: media relations, crisis management, employee communication, reputation management, community relations, and product/brand communication. Almost 40 percent also include a public affairs/government relations function and 25 percent reported including investor relations.[7]

Others in the United States have applied a communication theory approach in defining corporate communication as "a large number of specialized and general stakeholders in a number of important areas, including employees, customers, financial markets, and government regulations."[8]

Multinational corporations have influenced the discussion in Europe, where there are now undergraduate and graduate programs offering university degrees in corporate communication. Centers for Corporate Communication research in Denmark, Norway, and Amsterdam have taken a more theoretical approach to the discipline. New textbooks for European students have defined corporate communication from a more theoretical perspective:

> …combining theoretical and practitioner orientations will be advantageous in that it leads to theory and practice informing each other and ultimately will advance our understanding of the field of corporate communication as a whole.

> Corporate communication is a management function that offers a framework and vocabulary for the effective coordination of all means of communication with the overall purpose of establishing and maintaining favorable reputations with stakeholder groups upon which the organization is dependent.[9]

> Corporate communication can be described as the orchestration of all the instruments in the field of organizational identity (communication, symbols and behavior of organizational members) in such an attractive, realistic and truthful manner as to create or maintain a positive reputation for groups with which the organization has an interdependent relationship (often referred to as stakeholders). This results in a competitive advantage for the organization. Theoretically speaking, corporate communication can be divided into three main forms of communication: management communication, marketing communication and organizational communication.[10]

In 2006 the Commission on Public Relations Education issued an influential report. It documented the dramatic changes in the way corporations and organizations communicate both internally and externally with all of their audiences and constituencies. It recommended that graduate programs reflect the new focus on the discipline as a strategic management function. The PR leaders supported several types of graduate public relations programs rather than endorsing the MBA degree or dismissing public relations graduate education as unnecessary. They felt that graduate education should be interdisciplinary, combining public relations, communication and management courses.

> Research for the Commission's 2006 report included a review of public relations program graduate Web sites, telephone interviews with 18 public relations leaders and a quantitative survey of educators and practitioners.

> The Commission's quantitative study tested several options for graduate education. More than 60 percent of practitioners and educators sampled agreed that graduate public relations education should be an academic area of study with interdisciplinary focus (communication, management and behavioral science), or an academic area with a management focus. The survey participants' opinions reflected three different, but also overlapping, profiles that the Commission labeled the academic disciplinary focus, the academic focus and the professional focus. Few participants said that no graduate education was needed.

> In 2006, practitioner respondents in the Commission's qualitative survey recognized trends that were driving public relations that were not as important in 1999: rapidly changing new media; transparency and accountability demands; recognition of PR's increasing value by top management; the need and demand for measurement; globalization; diversity; ethics issues and credibility crises; more multi-disciplinary and integrated communication; and the need to align public relations with business strategy and social demographic changes. This research suggests that graduate education should move toward understanding business, management and public relations as strategic management functions.[11]

Considering the shift driven by globalization, media democratization (Web 2.0), and a transformed business model, the definition above (see page 15) serves professionals in Western, publicly traded companies well.

Strategic Communication Functions in Corporations

The communication role has changed in corporations just as the nature of the corporation has changed in response to an explosion of new communication technologies as well as global networks within organizations. Communication is more complex, strategic, and vital to the health of the organization than it was previously, and will only gain in importance in the information-driven economy. It is tied to the messages created for all audiences, internal and external, paying and non-paying. When the chief communication officers of publicly traded, multinational corporations were asked to rank descriptions of the leadership role of communication professionals in the corporation,[12] they responded with the following order of priority in terms of the percentage of respondents who ranked these functions first:

23.3%—Counsel to the CEO & the corporation [16.7% #2]
18.0%—Manager of company's reputation [14.8% #2]
12.9%—Source of public information about the company [14.5% #2]
10.2%—Driver of company publicity
9.7%—Manager of the company's image
6.8%—Advocate or "engineer of public opinion"
5.8%—Manager of relationships—co. & NON-customer constituencies
5.0%—Branding & brand perception steward
3.5%—Member of the strategic planning leadership team
3.4%—Manager of employee relations (internal comm.)
1.9%—Manager of relationships—co. & ALL key constituencies
1.8%—Support for marketing & sales
1.8%—Corporate philanthropy (citizenship) champion

The Functions of Corporate Communication

The CCI Corporate Communication Practices & Trends Studies ask communication executives whether or not their corporate communication responsibilities and budgets include more than 20 communication functions such as Annual Report, Crisis Communication, Employee Relations, Investor Relations, Internet, Intranet, Media Relations, Communication Policy, Communication Strategy, and Public Relations. Their responses, shown in Table 1.1, outline the core functions for corporate communication. The first eight, Media Relations, Public Relations, Communication Strategy, Crisis Communication, Communication Policy, Executive Communication, Reputation Management, and Employee Communication, are ubiquitous and form the core corporate communication functions.[13]

Table 1.1 Core Functional and Budget Responsibilities of Corporate Communication

FUNCTION	RESPONSIBILITY	BUDGET
Media Relations	100.0% *	86.2%
Public Relations	98.4% *	86.2%
Communication Strategy	96.9% *	76.9%
Crisis Communication	93.8% *	78.5%
Communication Policy	92.3%	69.2%
Executive Communication	87.7% *	73.4%
Reputation Management	84.6% *	73.8%
Employee Communication	81.5% *	75.4%
Social Media	78.0%	64.0%
Internet Communication	76.9%	63.1%
Intranet Communication	76.6%	66.2%
Annual Report	75.4%	63.1%
Corporate Identity	69.2%	56.9%
Issues Management	67.7%	55.4%
Community Relations	61.5%	47.7%
Mission Statement	56.9%	41.5%
Corporate Citizenship	50.8%	33.8%
Brand Strategy	50.8%	44.6%
Marketing Communication	41.5%	36.9%
Advertising	41.5%	35.4%
Corporate Culture	40.0%	29.2%
Investor Relations	32.3%	27.7%
Government Relations	15.4%	15.4%
Technical Communication	13.8%	10.8%
Ethics	9.2%	4.6%
Training & Development	7.7%	7.7%
Labor Relations	4.6%	3.1%

*Almost ubiquitous

Source: www.corporatecomm.org/studies

These figures indicate substantial involvement of corporate communication executives in communication actions central to corporate growth and survival. The responses also indicate substantial budgetary responsibility for traditional communication functions and a shared or matrix role in forging important corporate relationships with customers, vendors, and investors.

Corporate Communication Budgets of Publicly Traded, Multinational Companies

Figure 1.3 shows that the annual dollar value for corporate communication operations generally clusters around $1,000,000–$5,000,000 for about 35 percent of companies since 2005, and above $20,000,000 for 15 percent of corporations in 2009.

Budget changes from 2001 to 2009 shown in Figure 1.4 also indicate the value of corporate communication as a strategic management function. Almost 38 percent of corporate communication executives reported an increase in their budgets for 2007. With the 2008 global

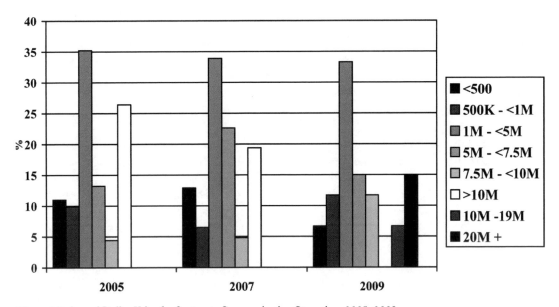

Figure 1.3 Annual Dollar Value for Corporate Communication Operations 2005–2009

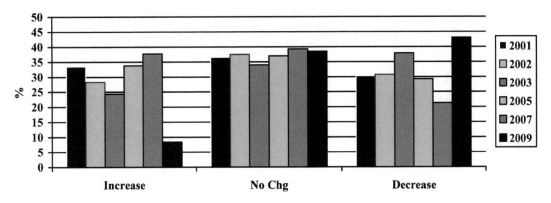

Figure 1.4 Corporate Communication Budget Changes 2001–2009

economic downturn, there were dramatic budget cuts: 43.1 percent reported more than a 5 percent decrease, contrasting sharply with modest increases in the past. Nevertheless, communication executives remain optimistic that budgets will not be "among the first to be cut" (90.8 percent), reflecting the value of the function.

Another study weighed the economic impact of the recession in 2008–2009:

> While the recession has certainly hurt, and there will undoubtedly be more pain in the future, our survey respondents, who come from a wide swath of the economy, have experienced significant but not debilitating budget cuts and have been able to prevent, at least through today, widespread layoffs. This is a significant change from the historical pattern, which saw precipitous cuts and sometimes near total elimination of PR/communication, in difficult economic times.
>
> The only plausible reason for this change is recognition that in our hyper-informational, increasingly transparent environment, organizations of all types need to communicate effectively or see their relationships with their key audiences wither away. This seems to be true even when—or perhaps especially when—times are tough. Engaging with your stakeholders in cost effective ways is no longer an optional practice; it's essential. While a sample of this type cannot be assumed to reflect the entire population of organizations, anecdotal evidence and our own experience in the field suggest that the results portray a reasonably accurate snapshot of the current situation.[14]

A benchmark paper in 2000 by the Council of Public Relations Firms Spending Study asked more detailed questions about spending in the Fortune 500. Some of the findings were:

> The "typical" corporate communication department in the study had a budget of $7.5 million and a staff of 10 professionals and three support staff. It was headed by a VP (often a senior or executive VP) who reported to the chief executive or chief operating officer, and expects next year's budget and staffing will both increase.
>
> The range of spending on corporate communication was very large: $285,000 to $100 million. The mean was $21.6 million.
>
> Among those companies whose budgets included them, the following were the largest line items:

- Corporate advertising ($11.4 million)
- Foundation funding ($8.1 million)
- Social responsibility ($4.65 million, including community relations, non-foundation funding, etc.)
- Government relations ($4.2 million)
- Employee communication ($2.6 million)
- Investor relations ($2.1 million)[15]

These figures underscore the importance of communication for Fortune 500-level corporations and the substantial financial commitment dedicated to corporate communication as a strategic management function.

Use of Vendors and Agencies in Corporate Communication

Corporate communication executives used vendors and agencies for their work. The most commonly cited were:

Advertising	63.1%
Annual Report	49.2%

Brand Strategy	41.5%
Media Relations	36.9%
Internet	35.4%
Public Relations	33.8%
Marketing Communication	33.8%
Government Relations	30.8%
Crisis Communication	29.2%
Social Media	28.0%
Identity	27.7%
Intranet	16.9%
Reputation Management	15.4%
Employee Communication	13.8%
Issues Management	13.8%
Investor Relations	12.3%

Source: www.corporatecomm.org/studies

Only 6.5 percent used a vendor for community relations, 5 percent used a vendor for labor relations, and 3 percent used a vendor for corporate citizenship activities. Fewer than 2 percent of companies used vendors for communication policy, corporate culture, and mission statement.

Investors, the community, vendors, and employees perceive the corporation through its actions. Its reputation is driven by how it behaves in these ways:

- How it runs its business—Governance
- How good a citizen and neighbor it is—Environmental and Social Performance. Even in a recession, people now expect such behavior.
- How good its products and services are—Brand Equity and Quality
- How its people behave—Human Capital
- How good its leaders and their plans are—Leadership and Strategy
- How good its vision for the future is—Innovation and R&D
- How the corporation behaves—individually and collectively—on a minute-to-minute basis with everyone, everywhere, and at all times is the basis for its reputation.

To meet the challenges of the forces of globalization, Web 2.0, and the Corporate Business Model: The Networked Enterprise demands enhanced leadership capabilities for both corporate communicators and the corporations they lead.

Notes

1. Facebook page, "Facebook: Stop Invading My Privacy!" http://www.facebook.com/group.php?gid=5930262681.
2. Thanks to two of Goodman's graduate students for their insights into IBM: Jenna Gable, MA in CC Baruch College, 2009; and Amy Paulsen, Corporate Reputation and Sustainability graduate seminar, Aarhus School of Business, Summer 2009. Also, see the IBM discussion at www.research.ibm.com/SocialComputing/WorldJam.htm.

3. George Soros, *The New Paradigm for Financial Markets* (New York: Public Affairs, 2008), 10.

4. Joseph A. Schumpeter, *Capitalism, Socialism and Democracy* (New York: Harper & Row, 1942), 82–85.

5. Arthur W. Page Society, *The Authentic Enterprise* (2007), 16.

6. Michael B. Goodman, "Corporate Communication Practice and Pedagogy at the Dawn of the New Millenium," *Corporate Communication: An International Journal* 11, no. 3 (2006): 197.

7. Paul Argenti, *Corporate Communication*, 4th ed. (New York: Irwin/McGraw-Hill, 2003).

8. Alan T. Belasen, *The Theory and Practice of Corporate Communication: A Competing Values Perspective* (Los Angeles and London: Sage, 2008), 8.

9. Joep Cornelissen, *Corporate Communication: Theory and Practice* (London: Sage, 2004), 11, 23.

10. Cees B.M. van Riel, "Defining Corporate Communication," in *Corporate Communication*, 2nd ed., eds. Peggy Simcic Bronn and Roberta Wiig Berg (Oslo: Gyldendal, 2005), 22.

11. "Graduate Education," in *The Professional Bond: Report of the Commission on Public Relations Education*, November 2006, 51–52; **www.commpred.org**.

12. CCI Studies 2009, www.corporatecomm.org/studies.

13. See www.corporatecomm.org/studies for reports on all the studies since 1999, including two China and one South Africa Benchmarks.

14. http://annenberg.usc.edu/AboutUs/News/090225SCPRCsurvey.aspx.

15. James Hutton and others, "Reputation Management: The New Face of Corporate Public Relations?" *Public Relations Review* 27 (2001): 247–61.

CHAPTER TWO

Leadership Capabilities

The Core Competencies for Corporations and Executives

Roger Bolton advised graduate-degree candidates at Baruch College/City University of New York that three characteristics are necessary for a corporate communication professional to succeed in the dynamically changing global environment described in Chapter 1:

- A clue—a deep understanding of the business the company is in; its business model for making money; how it aligns its mission with the demands of the marketplace and the community
- Guts—the ability to provide objective counsel to corporate officers, members of the board, and shareholders, even if that advice is unpopular
- "Woo"—like jazz, this term encompasses a complex set of behaviors including diplomacy, tact, personal credibility, "cool," and independent thinking, combined with a collaborative, team spirit

Roger has counseled presidents and CEOs. He has been the president of the Arthur W. Page Society. In his blog post "The Audacity of Authenticity,"[1] Bolton notes that President Barack Obama seems to be following the recommendations of *The Authentic Enterprise: An Arthur W. Page Society Report* (2007)[2] by articulating and activating shared values, building relationships among many groups of stakeholders, leveraging and empowering people with new media skills and tools, and building trust. *The Authentic Enterprise* defines the leadership role of the Chief Communication Officer with the following four priorities and capabilities:

1. defining and instilling company values;
2. building and managing multi-stakeholder relationships;
3. enabling the enterprise with "new media" skills and tools;
4. building and managing trust, in all its dimensions. (*The Authentic Exterprise*, 7)

This definition of the communication officer's role implies a two-way, mutual relationship between leaders and followers. Edwin P. Hollander, a recognized expert on leadership research and practice, identifies the 4Rs vital to successful relationships within companies—Respect, Recognition, Responsiveness, and Responsibility.[3] An inclusive approach to communication reflects the collaborative model of business that has in many global corporations replaced the command-and-control model.

The elevation of communication and its strategic importance to the health of the corporation is also reflected in a study of CEOs, senior strategists, and communication professionals that finds that "communication is unilaterally deemed critical to the success of strategic initiatives."[4]

CEOs look for the following capabilities and leadership qualities in an ideal chief communication officer:

Detailed knowledge of the business. This is far and away the most critical quality for a top communication executive. All CEOs believe that their businesses are large and complex entities, and that their companies cannot be communicated well if their top communication executives do not intimately understand them. CEOs also feel this hurdle is where some communication professionals fall short—where their knowledge is limited to communication and does not include knowledge of business in general and the details of their company's business in particular.

Extensive communication background. Experience and extensive relationships are to some extent taken for granted, although CEOs in different industries see nuances in the kind of background a given executive has. CEOs in highly regulated industries, for instance, are likely to put a particular premium on government and/or political experience. CEOs are clear, though, that while a strong communication background is necessary, it is not sufficient.

A crystal ball. CEOs say that in today's business context, a communication chief's ability to "see around corners" and anticipate how different audiences will react to different events, messages, and channels is critical. CEOs recognize that the proliferation of media and audiences means that the communication head's job is extremely demanding, but, in their minds, that's the price of success.

C-suite credibility. While CEOs acknowledge that credibility is subjective, they feel across the board that it's crucial for the communication head to be accepted at the highest levels. One particularly strong credential is experience in running a business or major division, whether at the present company or elsewhere.

Extensive internal relationships. CEOs want their communication head to have his or her finger on the pulse of the company, which means that this executive knows employees at every level of the operation.

Team player. A CEO's key decisions are generally made on a collaborative basis through a series of conversations with groups of people over time. Any serious player in these talks must have strong relationships with colleagues and the respect of the CEO's inner circle. Corporate and individual leadership is essential for positive and successful performance.

Educator. CEOs increasingly see the need for the communication head to educate the rest of the company on communication skills in general, as well as to generate strategies for communicating the company's values specifically. They also see building the internal network as a key asset for their communication head. (*The Authentic Enterprise*, 44–45).

Leadership in Corporate Communication

A simple key word search of "Leadership" on BarnesandNoble.com yielded 19,714 book titles in January 2009, down from 23,328 titles in January 2008 and 21,457 titles in 2007; and up from 19,150 in January 2006, 17,419 in 2005, and 15,323 in 2004. Many people have an opinion on leadership. "Leaders build bridges of hope and opportunity," says management development speaker Joel Barker. And many management experts consider that a leader's primary responsibility is to take care of tomorrow and to have a compelling vision of the future driven by an authentic and credible narrative, delivered in a clear and cogent way.

Some writers on the subject of leadership believe that leaders are born, while others believe that leaders develop. Still others believe in a combination of "in the right place at the right time." In his widely read book on successful business, *Good to Great,* Jim Collins[5] describes what he calls a Level 5 leader as someone who embodies a paradoxical mix of personal humility and professional will. Such persons are at once driven and humble. They set up successors because they see their corporation as a sustainable, long-term commitment. Because they know that they cannot possibly do all the necessary work by themselves, they are compellingly modest. And they are driven to produce sustained results because they think long-term and see beyond the horizon. Such leaders are workmanlike and diligent because they know that success depends on a thousand small actions done well, and done consistently. They know that their work is a team effort, and so they credit others with the successes of the group. And when things do not turn out as anticipated, they take responsibility for failure. Their humility drives them to attribute success to good luck, rather than personal effort.

Such leaders exhibit tough resolve in addition to humility. Companies that create an environment for such leaders to thrive, according to Collins, focus on people. They first assemble the best team they can, then address the specifics of the task. They first ask "who," then "what." They strive to enlist the right people on their team, and when an individual does not fit with the team, they are rigorous, rather than ruthless, in making decisions about people. Collins advocates three practical disciplines for rigor in people decisions:

- When in doubt, don't hire—keep looking.
- When you know you need to make a people change, act.
- Put your best people on your biggest opportunities, not problems.

In practice, such management teams debate issues and problems vigorously; then once a decision is made, they unify behind the decision. Dick Martin, in *Tough Calls* (2004), describes such decision-making when he was the Chief Communication Officer at AT&T.

As the model for global business has changed, the emphasis for many global enterprises has come to focus on innovation and ideas. Leaders look for people who exhibit the character traits and innate capabilities that underscore this new emphasis, whereas in the past people were recruited for their specific knowledge, background, and skills. Successful companies no longer pay lip service to the concept that the right people are their most important corporate asset.

Effective leaders are objective as well as optimistic. Collins says that they "confront the brutal facts, yet never lose faith." In an environment of ever-increasing transparency and disclosure for publicly traded companies, they make a good-faith, honest, and diligent effort to find the facts and articulate the truth about the company. Such transparency and honest assessment demand that leaders create a culture of accountability and candor.

These leaders are truthful with their employees and create an environment in which people and the truth can be heard. Such companies have learned from corporate scandals such as Enron that to create a climate of truth and candor, their leaders need to ask questions, because they do not assume that they know the answers. Because they believe in collaboration and respect for all employees, they foster spirited debate and healthy dialogue. They avoid coercion, because that undermines the culture they have established. When things do not go according to plan, they actively and objectively seek information that can lead to explanations and provide a path toward solutions, and they do this without seeking to assign blame in order to identify the lessons they have learned from the experience. To make sure that they hear the good news as well as the bad, they create reporting processes that clearly identify negative—as well as positive—information.

In answering the question posed by the title of their *Harvard Business Review* article "Why Should Anyone Be Led by You?"[6] British researchers Robert Goffee and Gareth Jones recognize that effective leaders possess vision, energy, authority, and strategic direction. Their research suggests four "unexpected" qualities of leaders:

- They selectively show their weaknesses. By exposing some vulnerability, they reveal their approachability and humanity.
- They rely heavily on intuition to gauge the appropriate timing and course of their actions. Their ability to collect and interpret soft data helps them know just when and how to act.
- They manage employees with something we call tough empathy. Inspirational leaders empathize passionately—and realistically—with people, and they care intensely about the work employees do.
- They reveal their differences. They capitalize on what's unique about themselves.

James Kouzes and Barry Posner's "Ten Commitments of Leadership" has become standard since it was introduced in the early 1990s in their *The Leadership Challenge*:[7]

Model the way—
1. Find your voice by clarifying your personal values
2. Set the example by aligning actions with shared values

Inspire a shared vision—
3. Envision the future by imagining exciting and ennobling possibilities
4. Enlist others in a common vision by aspiring to shared aspirations

Challenge the process—
5. Search for opportunities by seeking innovative ways to change, grow, and improve
6. Experiment and take risks by constantly generating small wins and learning from mistakes

Enable others to act—
7. Foster collaboration by promoting cooperative goals and building trust
8. Strengthen others by sharing power and discretion

Encourage the heart—
   9. Recognize contributions by showing appreciation for individual excellence
   10. Celebrate the values and victories by creating a spirit of community

The leadership challenge in the contemporary corporation is to build a communication culture from the CEO down to every member of the organization. The challenge is both on the individual level and the corporate level. That is, the people or personalities exhibit habits of behavior that lead them toward success; the organization as a whole also exhibits actions and behaviors that lead to success. Individuals exhibit leadership qualities that align with the corporate culture. We can think of corporate culture as the behavior of the organization as a whole that drives it toward success.

Individual leaders can transform the organization and take it into the future. Followers are primarily focused on the present in order to survive.

A profound connection exists between a leader's style and whether it leads to a positive or a negative outcome. A corporate leader's vision of the future has enormous power, and a shared vision between the leader and the members of the organization is a powerful component of success.

To adapt to a transforming global business environment, to Web 2.0 and 3.0, and to the emerging networked enterprise, this definition should prove useful: "A leader is someone you choose to follow to a place you would not go to by yourself" (Michael B. Goodman, CCI Leaders Forum, January 2001.)[8]

General Competencies for Global Communication Professionals

Competencies that are required to meet the demands of a global business environment include a combination of knowledge, skills, and attitudes that are appropriate to the context. Individuals need a set of core competencies to be able to learn and develop, to participate as active members of the society, and to contribute to the community as citizens and employees. The European Union identified eight competencies for successful lifelong learning:

- Communication in the mother tongue
- Communication in foreign languages
- Mathematical competence and basic competencies in science and technology
- Digital competence
- Learning to learn
- Social and civic competencies
- Sense of initiative and entrepreneurship
- Cultural awareness and expression[9]

The foundation for successful personal adaptation in the context of breathtaking global change is the ability to learn how to learn and is built on solid language skills, literacy, math and digital literacy, communication and information technologies. This foundation is fundamental to the ability to master the forces that shape the practice of corporate communication. Woven into these core competencies are critical thinking, creativity, innovativeness, problem solving, risk assessment, decision making, and emotional and social intelligence.

Chief Communication Officers, when asked, have strong convictions about what is needed to be successful in their role. Figure 2.1 lists the skills and personal characteristics, based on the results of the CCI Studies, necessary for success as a corporate communication professional. Certainly credibility is fundamental for effectiveness in an age of distrust and skepticism. It is important for communication professionals to develop and project these qualities and abilities.

• *Integrity & Honesty*	• *Resilience*
• *Global mindset*	• *Energy, discipline, passion*
• *Business orientation*	• *Leadership*
• *Critical & analytical thinking; synthesizing*	• *Intelligent; Innovative; Creative*
• *Communication & Media*	• *Emotional intelligence*
• *"Grace under pressure"— Confidence, composure, compassion*	• *Mentoring & Coaching; Quick study*
	• *Strategic thinking*

Figure 2.1 Skill Set for Corporate Communication Professionals from Responses to the CCI Studies Survey
Source: www.corporatecomm.org/studies

In the new business model—networked enterprise—the nature of work in the 21st century, as we have discussed above, is in a continuous state of change from the "command-and-control" models of the 19th and 20th centuries (Figure 2.2) to an "inform-and-influence" model of communication. To respond quickly, efficient corporations feature flat management structures with empowered managers. Power is distributed and people are networked with one another through corporate intranets. Command-and-control in this environment has given way to an inform-and-influence environment.

Work emphasis was on control:	*Work now focuses on influencing:*
• Mass market	• Customer-focused
• Mass production	• "Mini mills"
• Segmented	• Integrated
• Isolating	• Collaborative
• Mechanical	• Natural
• Mind numbing	• Challenging
• Hazardous	• Safe (safer)
• Repetitious	• Creative; vital
• Stressful	• "Fun"
• Disconnected	• Hyper-connected
• Translucent (unclear)	• Transparent
• Analog	• Digital
• Local	• Global

Figure 2.2 Changing Work Patterns and Attitudes toward Work Describe the New Networked Enterprise—From Command-and-Control to Inform-and-Influence

The process of work has changed as well. (See Figure 2.3 below.) Paradoxically, companies with strong corporate cultures and flat organizational structures often invoke more micromanagement through reporting and bureaucratic procedures. Middle managers are often considered a communication bottleneck in such lean organizations. Communication networks and strategic brainstorming "Jams" flatten the organization even more.

Work was:	*Work is now or is becoming:*
▪ Centralized	▪ Decentralized
▪ Functional groups	▪ Project team-oriented
▪ Assembly lines	▪ Empowered group
▪ Compartmentalized	▪ Multi-disciplinary
▪ Fragmented (task-oriented)	▪ Results-driven
▪ Stressful (physical & mental)	▪ Fulfilling
▪ Dictatorship	▪ Democratic
▪ Protective	▪ Creative
▪ Militaristic	▪ Collaborative

Figure 2.3 Impact of Change on the Process of Work

Another paradox has developed. Attitudes toward work underscore the volatility of employment and the decades-long message from corporations that job security cannot be guaranteed. So, in a decentralized, empowered, and electronically mediated work environment, employee loyalty is valued, but not expected. In a global environment, professionals work with colleagues they may know only through electronic messages. In recognizing that people still need to meet to build relationships, many companies create working groups that have initial meetings and will often meet periodically to reinforce those connections.[10]

The Career Map for Corporate Communication Professionals

Corporations consider the development and performance of their communication professionals to follow, in general, five core career levels, as shown in Figure 2.4: entry professional, professional, senior professional, executive (operating company), executive (corporate level). The generalist's path applies to all five levels. Specialist paths—media relations, marketing communication, organizational communication, emerging media—follow a path from the learning level, through applying, to leading.

Of course, as the models and patterns of business change and companies flatten their management structures, the career paths that worked well for individual development in a more hierarchical business model have given way in the networked enterprise to more detours to consider, as described by Wyeth's corporate communication executive Wendy Kouba and shown in Figure 2.5. For Kouba, the detours along the way are extraordinary opportunities for personal and professional development.

	Career Levels / Career Paths	Generalist Path	*Media Relations* Alternate Specialist Path	*Marketing Communications* Alternate Specialist Path	*Operational Communications* Alternate Specialist Path	*Emerging Media* Alternate Specialist Path
Expert	*Executive Vice President Operating Company, Corporate or Franchise level*	X				
Leading	**Senior Professional** Sr. Director Director	X	X	X	X	X
Applying	**Professional** Manager	X	X	X	X	X
Learning	**Entry Professional** Sr.Specialist Specialist Associate Coordinator	X	X	X	X	X

Figure 2.4 Five Core Communication Career Levels and Paths
Source: Wendy Kouba

Corporate Functions that Offer Professional Development Opportunities	
• Business Development • Change Management/ Organizational Development • Community Relations/ CSR/Philanthropy • Diversity • Federal/State Government Relations • International Assignments • Investor Relations	• Merger & Acquisition Integration • Public Policy/Public Affairs • Sales & Marketing • Strategic Planning • Manufacturing • Consulting • Corporate, Site & Sector Roles

Figure 2.5 Corporate Communication, Unlike Other Disciplines, Can Offer Valuable Detours to Consider along the Career Path
Source: Wendy Kouba

The core competencies for successful individual development in corporate communication focus on:

Technical Skills. Demonstrate strength in written and oral communication, as well as the ability to obtain information, critique, edit, and communicate effectively.

Product and Market Knowledge. Shows an understanding of the customer, product, and market to effectively develop and deliver communication.

Business Leadership. Demonstrates knowledge of business complexities and shows an ability to act as a strategic partner to leverage the power of communication to drive business.

The Industry Sector Environment. Demonstrates knowledge of how the company makes money, as well as related legal, regulatory, and policy requirements.

Reputation Management. Demonstrates the ability to build public trust and enhance the reputation of the business.

The core competencies for specialists in these four areas focus on:

Organizational Communication. Understands business strategy. Leverages internal communication process to convey messages and influence employees to inform, engage, and align.

Marketing Communication. Understands how to integrate diverse media channels into overall communication mix to effectively promote products, services, and organizations.

External Communication. Understands the importance of maintaining a good corporate image and exhibits skill in interacting with media regarding products, promotions, and issues.

Emerging Media. Demonstrates basic skills in leveraging emerging technologies to advance communication.

Because it is a flat, collaborative organizational structure, "soft skills" matter in the networked enterprise. The individual must recognize, then, the value that he or she brings to the table. Effective professionals allow people to get to know them. They build relationships at all levels, and they cultivate "360° supporters." (See Figure 2.6.) They learn to understand—and value— others' styles. They build their own self-awareness and ask others what they need. They are always mindful of the context and environment of their interactions. They know what they know and what they don't, and they ask good questions.

In building a career in corporate communication, Kouba advises that an individual must take responsibility and manage it, and that it is important to know that there is no "right" career path. She notes that generalists tend to be more marketable and resilient when they have depth in key areas. Experience in other industries is a plus, since it demonstrates an ability to assimilate information quickly. Global experience is critical, and mastery of multiple languages is a distinct advantage. She advises professionals to develop and continually refine leadership. Also, she suggests that it is necessary to be cognizant of the business environment and to learn to thrive on ambiguity. Finally, in response to continuous change, she urges that professionals be able to reinvent themselves when necessary.

Aspire High *Set high goals for self and others and inspire high performance.*		Think Broadly *Operate with the customer in mind and reach beyond functional boundaries to meet stakeholder needs.*
	Execute Flawlessly *Get things done and deliver quality results.*	
Be Decisive *Make and communicate high-quality, timely decisions that deliver superior performance.*		Build Talent *Attract, engage, and develop people to outperform the best in the industry.*

Figure 2.6 Competencies in the Corporate Context for Building Successful Internal Relationships

The Key Functions of Corporate Communication Indicate Leading Practices

Insights from the CCI Studies outline opportunities for future practice, professional development, and development of a curriculum for higher education.[11]

The need to build trust with all internal and external audiences. In their comments and in the interviews, corporate communication executives indicated a clear need to build trust with all audiences. This has been true since the 2002 study. In practice this means that the corporation's relationships with external and internal communities matter a great deal. Customers, vendors, and business partners need a firm relationship of trust in the wake of the 9/11 attacks on the World Trade Center, the accounting scandals that followed in the fall of 2001, and the erosion of confidence in the capital markets. "Formal trust," as Goodman noted in the *Journal of Business Strategy* 26, no. 4 (2004), includes the rule of law, transparency, and publicly evident rules. Informal trust is culturally defined by the values and norms that allow people to communicate and deal with others who share those values.

"Trust is the key objective for global companies today because it underpins corporate reputation and gives them license to operate," said Michael Deaver, Vice Chairman of Edelman, in a note about the Edelman Annual Trust Barometer presented in January 2006 at Davos. "To build trust, companies need to localize communication, be transparent, and engage multiple stakeholders continuously as advocates across a broad array of communication channels."

The business case is a simple one: the license to operate is either granted, or revoked, by the society one is in. This concept is clear for companies in the European Union and the United Kingdom, which emphasize sustainability, or the triple bottom line—financial, environmental, and social performance measures. The trend has become a leading practice among global corporations: engage the public, customers, employees, and business partners on the social, financial, and environmental accomplishments and actions of the corporation. A new era of transparency has created an opportunity for building trust through strategic corporate communication initiatives.

The expectation by the corporation to accomplish more with less. Even though budgets for corporate communication increase slowly, staff responsibilities increase more rapidly. So following a trend that CCI research has noted since 2000, and what one communication officer described as "always more to do," professionals will continue to be asked to be more productive as companies continue to expect corporate communicators to accomplish more with fewer resources. In this constant escalation of the volume of work, the opportunity is there for professional development and cross-fertilization of skills in order to make each member of a corporate communication staff more capable in all functional corporate communication areas. The challenge for communication officers is to develop measures for corporate communication value. As their corporations turn to various measures, several are developing corporate communication criteria to fit with the "balanced score-card" that their corporations use to judge performance.

The demand to build a responsible and accountable global corporate culture in response to a hostile environment for multinational corporations. Anti-Americanism and anti-globalism create a hostile environment for multinational corporations, making daily operations more challenging than ever. The trend toward building a responsible and accountable global corporate culture is vital to organizational health. Corporations that operate globally work hard to harmonize their corporate cultures with the local culture. Look to companies that have operated globally for decades, some for more than 100 years, for best practices in creating an effective global corporate culture. The opportunity and challenge is for the corporation to address and embrace global issues for competitive advantage, and for multinationals to engage in "Public Diplomacy" to create a stable and peaceful business environment.

The perception of the corporate communication executive as "counsel to the CEO" and "manager of the company's reputation." Communication is now more strategic than ever. Corporate communication executives see "counsel to the CEO" and "manager of the company's reputation" as almost equal to their primary role in the company. A large majority have a seat at the table, with approximately 75 percent of communication officers reporting either to the CEO (53.8 percent), the CFO (14 percent), the COO (3.2 percent), or the Corporate Counsel (5.4 percent). In this role, the trend toward strategic communication counsel for the corporation accompanies the role of "Chief Communication Officer" (2001, 2002, 2003, 2005, 2007, 2009).

However, a counter trend—a business model that is worth watching—is emerging. The model involves a CEO who sees little value added in communicating with any constituency other than a handful of key investors. The role of the communication officer is then greatly diminished as some corporations merge, or as corporate communication executives retire and the position is restructured or fragmented. Essential functional responsibilities, such as crisis communication and media relations, or legally mandated financial reports, have been decentralized. For the majority of corporate communication executives, such responsibilities for corporate decision-making come with great opportunities and challenges to become trusted advisors to the corporation and its leaders.

The understanding of the global impact of the local act, and the local impact of the global act. Global issues—anti-globalism, outsourcing, terrorism—place corporate communication executives as strategic counsel to the corporation. Often the success of an offshore initiative is directly attributed to intangibles such as corporate culture or corporate communication. The trend is for corporate communicators to understand the global impact of the local act, and the local impact of the global act. The opportunity and challenge is to move from preaching "global" to embracing it with stories and messages.

A special issue on Public Diplomacy of the *Journal of Business Strategy* (June 2006) offers this observation:

> Contemporary business can be a powerful force for global change. A consensus is building that it is proper, and in some cases preferable, for business to marshal its resources to solve problems on a global stage. Business has a considerable role to play in Public Diplomacy, which was defined in 1965 by diplomat Edmund Guillion, and later Dean of the Fletcher School of Law and Diplomacy at Tufts University, as dealing with the influence of public attitudes on the forming and carrying out of foreign policy. And it is an activity in international relations beyond traditional political diplomacy that includes actions by governments to "cultivate" public opinion in other countries and to communicate with foreign correspondents. It is also the country to country interaction and intercultural communication of private groups such as businesses. Substantial agreement now exists that global business can and must act to solve problems that have often been handled by government. In the decades following the end of the Cold War government systematically backed away from issues of economic development, as the "conflicts" at the core of the aftermath of World War II seemed to evaporate along with the fall of the Soviet Union. However, the economic and social issues that were masked during the ideological conflict of the Cold War surfaced, as did the suppressed desires for political and social ambition.[12]

Executive compensation continues to be a controversial issue. Executive compensation issues have increased the need to clearly articulate "pay-for-performance" criteria, to educate constituent groups about retaining top talent, and to educate executives about the need for open, factual, non-defensive discussion of their compensation. Media and public awareness of the issue demands a greater concern for transparency, employee dissatisfaction, and stakeholder backlash. The "say-on-pay" proxy initiatives have gathered support of investors and analysts in the wake of the financial crisis of the fall of 2008, particularly the excessive bonuses some companies awarded their executives even as corporate performance indicated failure. For corporate communication professionals, the opportunity is to embrace through action and behavior the company commitment to accountability and transparency. (See the Guideline "Transparency and Disclosure" in Part 5 below.)

The demand for greater transparency and disclosure has made media relations more complex and strategic. News media and technologies are pervasive, instantaneous, and global. More news outlets demand more attention and new tools and techniques to meet the challenges of blogs, really simple syndication and consumer-generated media. Transparency and disclosure laws and practices have made media relations both more complex, and more strategic. One executive says the answer is to "become a 'glass-box' company." Since the trend is toward more, not less, disclosure of information, more demand for transparency should be expected. The opportunity for success lies in how soon a corporation can embrace new communication tools and technologies. The challenge for professionals is how to manage the messages.

The expectation that the company be a good citizen and make money. The community a company is in expects a strong and open dialogue about what the corporation believes and what it values. A strong relationship of trust closes the gap between perceptions of the company and its performance. The trend is that good corporate citizenship is expected. The opportunity for success is in building a strong reservoir of trust: positive reputation equity with all audiences. Business has rediscovered its purpose in this context as global citizen, expanding beyond Nobel Prize-winning economist Milton Friedman's definition more than three decades ago that the purpose of business is to create wealth for its owners, within the rules. The understanding of the meaning of "owners" has expanded to include:

- Non-governmental organizations (NGOs) that are also investors
- investors who are also employees
- employees who are also customers
- consumers who are also local business partners
- business partners who are also local stakeholders
- local stakeholders who are also media
- consumers who are also media
- media who are also NGOs

All business, as legendary AT&T executive Arthur W. Page observed more than 75 years ago, begins with public permission and exists by public approval. The expectation is that a company's enlightened self-interest will also cause it to contribute to the social good within its own areas of competency at the intersection of society's needs and its own interests.

The intangible risks of not acting as a good corporate citizen and public diplomat can be numerous. For example, it can result in poor social and environmental performance. It can often feed into an obsession with short-term financial performance. It can be the rationale for a lack of transparency in corporate reporting and a lack of stakeholder engagement. It can lead to an insular perception of the world outside the corporation that can result in limited risk management of critical issues.

Poor corporate behavior can also put the company's reputation at risk. Lawsuits (domestic and foreign) can result. Negative media coverage can be the outcome of poor media relations. NGO pressure and consumer boycotts can tarnish perceptions of the company and lead to reputation damage. Bad corporate behavior and opaque corporate communications can lead to the eroding of public trust and negative analyst assessments. Market punishment can also result.

The public acceptance of the role of business-as-citizen reminds us that corporations must act as good citizens or lose their license to operate.

The reality of global conflict makes crisis communication planning a critical success factor for corporate communication professionals. Companies will have unexpected—indeed inconceivable—crises, and preparations should be made for them. The trend toward increased global conflicts and economic shockwaves makes crisis communication planning a critical factor for corporate communication professionals. The opportunity to add value to the company is through continual crisis planning as well as professional development and cross-functional training through quarterly "crisis drills."

The understanding of transparency as a best practice strategy for corporate reputation management. Reputation is an intangible and valuable corporate asset. It is difficult to achieve a positive reputation, but even more difficult to protect it. As Baruch Lev and others have observed, more than 50 percent of the market value of a business can be attributed to intangibles. Indeed, institutional investors now make a substantial percentage of their decisions based on the intangibles they see in a corporation.

These intangibles drive corporate value: corporate reputation, governance, innovation, research and development, environmental and social performance, brand equity, human capital, leadership and strategy, product and service quality. Transparency is a leading practice strategy for reputation management. The opportunity for corporate communicators is to add value by offering constructive suggestions on improving corporate reporting and corporate governance.

Although writing is still the core skill for Corporate Communication, the communication officer needs to have business knowledge, to understand budgets, and to know how to measure the value of communication to the organization. As it has been since the first CCI study when interviewees were asked to identify the skills and capabilities needed for corporate communication, respondents identify writing as the core skill for corporate communication. Another essential skill is a thorough knowledge of the company and of business principles.

Additional Capabilities for Corporate Communication Professionals

In addition to writing expertise and superior interpersonal skill, the ability to create media products—press releases, video, web pages, magazine articles, newsletters, speeches, and blogs—is the ability to teach, to absorb and comprehend vast amounts of complex information quickly, to create and build relationships internally and externally, to build trust in audiences, and to build a corporate culture.

Teach

The corporate communicator is taking over a human resources role when it comes to motivating employees, and that calls for being a teacher. To do that requires awareness of the styles of adult learners, and we can begin by defining "adult" for many tasks as anyone over thirteen. Experience is key to adult learning. Almost everyone knows of someone who learned to program a computer by himself or herself, to fix an automobile or stereo system, or to have children and raise a family without the aid of formal schooling.

Absorb Complex Information Quickly

In times of challenge—crises, emergencies, mergers, acquisitions, or strikes—the appetite for quality information by one's community and one's corporate community is voracious. Stacks of information about the merger must be translated from "lawyer-" and "accountant-speak" to a language reasonably intelligent people understand.

People with a background in the Liberal Arts developed a facility for complex ideas and information when they interpreted the novels of Dickens and Twain, or understood the monographs of Freud and Jung, or decoded the field studies of Mead, or commented on the observations of Churchill, or criticized the thoughts of Socrates and Confucius, or misread the essays of Derrida. These skills are fundamental when translating complex ideas into everyday language.

Create and Build Relationships

Building relationships has never been easy, but in simpler times the role of employee communication was given to trusted employees who knew most everyone in the company because he or she had worked there since high school. In a "family" culture model, trust was institutionalized, since everyone in the company was family.

Relationships with the media were also built by recruiting reporters who had covered the company and by hiring them to be the media relations representative. With the change to a corporate community model, it is important to work at identifying the people who are essential to the organization and to cultivate a relationship with them.

Build Trust

How does one go about building trust? It is as simple—and as emotionally difficult—as being worthy of trust. People must have a sense of one's integrity as the cornerstone of trust. Every positive interaction with people builds a reservoir of trust. It is cumulative, from simply showing up on time, to keeping promises. On the other hand, integrity is something that can be lost only once.

Build Corporate Culture

When your employer asks you to take a few days to change the culture of the company, stop and take a few deep breaths before answering. You might ask if he or she has ever been swimming in the ocean and gotten caught in the receding tide, or felt the enormous pull of a rip tide. Or you might ask if he or she has ever been tossed around by a sudden gust of wind—a hurricane, or a tornado.

Culture is like the wind and the tide. It exerts very strong and often invisible forces. It has to be considered. Cultures can be changed, but not instantaneously. And rapid culture changes are painful and destructive, like wars. One doesn't need to look too far afield to find destructive cultural changes that are slow to heal—Ireland, the Middle East, Bosnia, Korea, or Central Africa. The citizens of the state of Georgia still recall with anger the slash-and-burn policy of Sherman during the Civil War. The British have no monument at Edgehill, the site of the first battle of their Civil War to overthrow the king by Cromwell and the Roundheads over 350 years ago.

Changes in Audiences and Communication Channels

Gen-X, Gen-Y, Gen-Z, the digital natives, Millennials….Every August for over a decade, Beloit College's Keefer Professor of the Humanities Tom McBride and Public Affairs Director

Ron Nief have developed a "Mindset List" for college professors to better understand "cultural touchstones that shape the lives of students entering college." In 2008 it noted:

> The class of 2012 has grown up in an era where computers and rapid communication are the norm, and colleges no longer trumpet the fact that residence halls are "wired" and equipped with the latest hardware. These students will hardly recognize the availability of telephones in their rooms since they have seldom utilized landlines during their adolescence. They will continue to live on their cell phones and communicate via texting. Roommates, few of whom have ever shared a bedroom, have already checked out each other on Facebook where they have shared their most personal thoughts with the whole world.[13]

Developing and maintaining the organization's culture has added to the challenge of corporate communication. Employees are no longer captives of the organization. They move often from job to job. They learned their lesson well from the experience of the decades of downsizing, restructuring, and mergers and acquisitions. They were told in school and observed from their parents that corporations and organizations would not have a job for them for life. They were taught in high school and college to see each job as a learning experience in preparation for the next job in their career path. Service was self-service, so they have no role models for understanding the concept of the value-added nature of customer relations. Enlightened self-interest was the appropriate way to think about their place in the world of work. They saw what happened to their fathers and mothers who committed themselves to work and a life of delayed gratification—downsized at age 55, just short of their pension and other benefits.

Now the challenge is to motivate a generation of workers who have priorities vastly different from the priorities of the company. Work/life balance for corporation places work first. For the new workforce, work/life balance means *life balance*. When work does not fit, it's time to move on. The low unemployment rate contributes to the validity of workers' approach to work.

Culture is also an essential understanding for the global workforce. Employers may understand the style preferences of their employees because they have their Myers-Briggs profiles. But an understanding of the influence of cultures from an anthropological perspective is also essential.

Cognitive psychologists are challenging the assumptions about our individual habits of thought. They are questioning the belief that "the strategies people adopted in processing information and making sense of the world around them…are the same for everyone…[that is] a devotion to logical reasoning, a penchant for categorization and an urge to understand situations and events in linear terms of cause and effect."[14]

Professionals from countries such as Japan, China, and Korea think "holistically." At the risk of gross generalization, they construct the world differently than Westerners do by paying more attention to context and relationships. Many also have an ability to hold contradictory thoughts simultaneously—Yin/Yang. So the audience analysis challenge becomes even more complex. Interestingly, Easterners born and raised in a Western environment show no clear preference for either rational or holistic thought as a result of strong competing cultural influences.

When All Is Said and Done

We conclude with nine personal and corporate goals to guide in strategic adaptation. The amount of information relevant to corporate communication requires (1) tenacious pursuit of intellectual competence in the field. The daunting skepticism about corporate business practices demands (2) constant and consistent demonstration of ethical understanding and awareness. The complexity of operating in a multinational business environment with numerous constituencies calls for (3) professional expertise and familiarity with research tools and techniques.

Rapid changes in technology and in global business practices requires (4) creative strategic integration of knowledge to "connect the dots," to see the patterns that others do not. The growing perception of corporate communication professionals as counsel to the CEO and the corporation suggests (5) the nurturing of leadership capabilities—specifically, a leader is someone who will take a person to a place he or she would not go otherwise. The transformational impact of social media, Web 2.0, and the semantic Internet, Web 3.0, requires corporate communication professionals (6) to cultivate media and technology literacy and expertise.

The corporate network capability demands that companies and their communication professionals (7) be global through strategic intercultural and global communication. As counsel to the corporation and part of the management team responsible for meeting the challenges it faces, a corporate communication professional must (8) be a problem solver through strategic decision-making, every minute and on every task. And, finally, as the voice of the corporation, a successful practitioner must (9) attain advocacy competence to become the best advocate for the company.

Authentic behavior provides a path for successful strategic adaptation.

With all the changes in the nature of work in the networked enterprise and the new tools to use in Web 2.0 and Web 3.0, people, working in a global environment, might find some simple guidelines helpful. Nordstrom's has two: 1. Use your best judgment; 2. See rule 1.

Judgment, wisdom, understanding, integrity—develop and rely on them.

Notes

1. Roger Bolton, "The Audacity of Authenticity," www.awpagesociety.com/awp_blog/comments/the_audacity _of_authneticity, January 25, 2009.
2. *www.awpagesociety.com/images/uploads/2007AuthenticEnterprise.pdf.*
3. Edwin P. Hollander, *Inclusive Leadership* (New York: Routledge, 2009).
4. "The Powerful Convergence of Strategy, Leadership, and Communications," Breakfast Briefing—Council of Public Relations Firms and the Arthur Page Society, New York City, July 2009.
5. Jim Collins, *Good to Great* (New York: HarperCollins, 2001).
6. Robert Goffee and Gareth Jones, "Why Should Anyone Be Led by You?," *Harvard Business Review* (September–October 2000).
7. James Kouzes and Barry Posner, *The Leadership Challenge*, 3rd ed. (San Francisco: Jossey-Bass, 2002).
8. Michael B. Goodman, CCI Leaders Forum, January 2001.
9. "Key Competences for Lifelong Learning: European Reference Framework" (Luxembourg: Office for Official Publication of the European Communities, 2007).
10. Stacey Connaughton, "Distanced Leadership," www.corporatecomm.org/archive.html, presented February 19, 2004.
11. The discussion that follows is adapted from Michael B. Goodman's "Corporate Communication Practice and Pedagogy at the Dawn of the New Millennium," *Corporate Communications: An International Journal* 11, no. 3 (2006): 196–213.
12. Special Issue on Public Diplomacy, *Journal of Business Strategy* (June 2006).
13. www.beloit.edu/mindset.
14. Erica Goode, "How Culture Molds Habits of Thought," *The New York Times*, August 8, 2000, D1, D4.

PART TWO

Understanding the Forces that Shift the Context of Corporate Communication

In Part One, we laid the foundation for the discussion of strategic adaptations in corporate communication by describing the geopolitical, economic, business, and media environments that have shaped the need for those adaptations. We also looked at the core competencies and leadership capabilities that have emerged as a consequence of the global shifts we have described.

In Part 2, we examine how the corporate communication challenges created by these shifts are beginning to play out by looking in more specific detail at:

1. How Web 2.0 is re-shaping four of the key corporate communication specialties
2. How changes in the business and geopolitical environment have impacted the ways in which corporations establish and maintain relationships of trust and integrity with stakeholders
3. What impact the underlying forces for change have had on corporate culture and its increased significance in the new environment
4. The new corporate communication needs created by economic factors in the changing world marketplace

In describing the way in which Web 2.0 is re-shaping corporate communication, we identify seven distinct features of Internet-based communication. By way of example, we then examine the impact of these new features on investor relations, on employee communications, on public affairs (government relations), and on community relations.

The impact of these features and the increasing fragility it can cause in stakeholder relationships serves as a suitable introduction to the issues of trust and integrity that have become such a prevalent component of the corporate communication ecosystem. In order to do so, we review the evolution of key corporate events since the beginning of the 21st century, including

Enron/Andersen and Sarbanes-Oxley, the regulatory response to incidents of accounting fraud. We also examine other changes in corporate governance that have impacted corporate communication in this millennium, as well as the significant growth in the corporate social responsibility movement as a response to popular concerns about the social role of the for-profit institution.

We conclude Part 2 by delving more deeply into the corporate communication ramifications of more recent macroeconomic factors, such as the rise of the BRIC countries (Brazil, Russia, India, and China) and how the shifting forces of economic nationalism conspire to create an environment of persistent instability to which the corporate communication function must address itself. These concluding observations about the global scene form an appropriate bridge to an examination in Part 3 of the strategic and tactical models available to the global communicator for managing the issues created by the web-enabled integrated global economy.

CHAPTER THREE

Corporate Communication and Web 2.0

The Internet has had a transformative influence in corporate communication from its beginnings to its current form as Web 2.0. That influence extends not only to the introduction of a wide array of new communications channels, but also to the very core of what we consider to be corporate communication. The extraordinarily high levels of interactivity and transparency enabled by the Internet have made the elemental practices of corporate communication (corporate reputation, employee communication, shareholder communication, community relations, and public affairs) unrecognizable to earlier practitioners.

The current transformation will continue, and it is likely that we will look back on this era as having changed not only the way companies communicate with these stakeholders, but the very nature of those relationships as well. We will be able to say in a few years that "the medium is the relationship," contrary to Marshall McLuhan's classic observation that "the medium is the message." Or perhaps we will adopt the concept proposed by Rich Teplitsky, head of the Public Relations Society of America's (PRSA) Technology Section, that "there are no more mediums, only messages."

The Growth of the Internet

With more than a billion Internet users worldwide, or about 20 percent of the world's population, the Internet dwarfs any previous human communications medium. The amount of Internet traffic continues to grow and will do so as the access to broadband penetrates more countries around the world. In July 2008, search provider Google estimated that it had identified one trillion indexed pages on the web. While the demographics of Internet users still show a skew toward younger age groups, by 2008 the medium had become almost age and gender agnostic. According to the Pew Internet & American Life Project, only 26 percent of Americans age 70–75 were online in 2005. By 2009, that number had risen to 45 percent.

While the increase in Internet accessibility is clearly leveling off in the developed world, continuing growth elsewhere and the sheer scale of human involvement with this new technology is remarkable. See Figure 3.1 below.

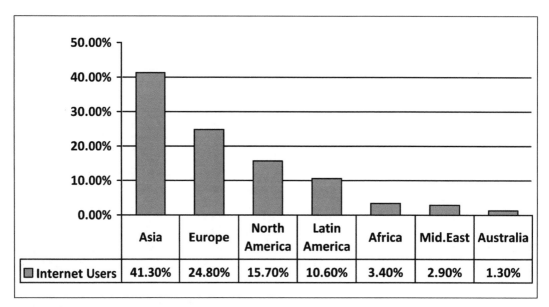

Figure 3.1 Internet Users by Global Regions
Source: Internet World Stats[1]

In addition, the reasons for Internet use and the "life segments" in which this use took place have also changed radically in the last decade. Whereas the consumer Internet's origins in multi-user dungeons and chat rooms made surfing the web a home or leisure activity in its early days, more than 50 percent of workers in large corporations report (Pew 2008) that they use the Internet at work constantly or several times a day. This number is even higher (72 percent) for government workers. Forty percent of American workers with PDAs also use their devices to check email on weekends.

However, research also shows that as the percentage of Americans with heavy work-related Internet usage increases, there remains a very substantial core of workers who never use the Internet for work at all. Fifty-two percent of service workers and 32 percent of workers in large corporations report no Internet usage at all in the workplace. This growing divide has immense ramifications for employee communications as the digital model of communication comes to predominate.

The Communication Impact of This Shift

The nature of the shift to Internet-based communications seems so pervasive and all-encompassing that it can seem daunting to try to identify the salient features of this shift independently. However, we can isolate seven distinct aspects of this revolution, each of them having profound significance to strategic communication. They are:

> *Multi-modal.* Web 2.0 delivers communication in text, audio, visual form, data and spreadsheets, video and widget format, and responses can also take any audiovisual form.
>
> *Time-agnostic.* Information can be delivered in real time, delayed time, or sampled as needed by the recipient.
>
> *Platform-independent.* The nature of the communication is not dependent on the delivery device involved. Mobile devices such as cell phones, PDAs, and laptops, as well as fixed devices such as desktops, display screens, and elevator panels are all equally available to digital content.
>
> *Automated.* With increasingly sophisticated tagging, search, and storage, information can be disbursed unprompted in response to search, delivered through alerts, RSS feeds, or activated through link streams.
>
> *Transparent.* Blogs, social media, and other platforms such as Twitter create "contact portfolios," making communications immediately transparent to multiple users. In addition, comment and response systems inside these platforms are visible to all participants.
>
> *User-generated.* Workplace wikis, polling systems, and prediction engines are the tip of the iceberg in user-generated content developed by stakeholders, from employees to customers.
>
> *Accretive/searchable.* The use of social media platforms breaks down barriers between media and allows for information/communication on a given topic to create accumulated searchable knowledge.

Each of these seven variables provides exciting new opportunities for communication but also creates new pitfalls for organizations trying to control their messages. In the following sections, we explore four important areas in which communication with varying stakeholders is influenced by the forces we have cited. These include investor relations, communications with institutional and individual shareholders; employee communications; community relations, the conversation with the communities in which companies have their facilities; and public affairs, the dialogue with legislators and regulators at a local, state (province), national, or international level.

Investor Relations Embraces Strategic Communication Technology

Both the most dynamic and yet the most conservative of stakeholder relationships, investor relations, has been transformed perhaps more than any other area of corporate communication by the advent of Web 2.0. In the early days of the Internet, small technology companies with enthusiasm for and understanding of the medium were among the first to use their websites to communicate with investors. However, large multinationals were not slow to follow, and by the late 1990s most public corporations were using the web as a repository for data they were required by law to share. These included press releases, annual reports, 10Ks, 10Qs, information about the election of directors, and data about the sale and purchase of company shares by officers and other corporate insiders.

The growth of the Internet also coincided with the movement, initially sponsored by non-governmental organizations (NGOs) but subsequently taken up by activist shareholders, to force corporations into greater disclosure and accountability. As the ideas of corporate social responsibility took hold, companies began to use the Internet to report on their steps to be better stewards of the environment, to describe their efforts to improve working conditions in their plants in the developing world, and to document their adherence to the principles of good corporate governance. In fact, corporate websites proved to be a useful repository for all of the policy positions and statements that companies felt the need (or were forced by shareholders) to communicate.

Out of this first wave, in which the corporation's increasing need for transparency met the evolving sophistication of the Internet, came the earliest attempts to use broadband to reach investors in a more visceral way. Cisco Systems was one of the pioneers of the use of webcasts to reach investors both on the day of a quarterly earnings announcement and as a stored resource. Combining audio of the analyst call with a visual click-through of the presentation itself democratized investor relations in a way that gave the average investor as much access as professional investment analysts once enjoyed.

In addition to augmenting traditional print media with audiovisual information, the digital Internet also transformed traditional print navigation. Whereas the reader of the Intel annual report would once have had to flip back and forth between sections, html now enables her or him to go straight from the consolidated statement of income to a footnote about employee executive incentive plans with one mouse click. This use of links has enabled companies to make available a vast library of investor resources without overwhelming the interface. With one click, investors can go to "Frequently Asked Questions" (FAQs), glossaries of investment terminology, deeper explanations of products or technologies, and even government databases of relevant compliance information.

In the attempt to create a sense of ownership and convenience for the investor, corporations such as Cisco, Walgreens, and Newell Rubbermaid have created a "briefcase" option so that an individual investor can aggregate his or her own documents of interest in a single place on the company's website. Almost universal in web-based communication at this point is the deployment of RSS (Really Simple Syndication) or investor alerts. By clicking on a link on most investor pages, the investor can request that breaking news and up-to-date financial information be sent automatically to a desktop, Personal Digital Assistant (PDA), or cell phone.

XBRL and Its Impact

Leaders in investor communications have worked hard to reduce the click stream required to get to different parts of investor pages, as well as to provide access to thought leadership presentations in a variety of audiovisual formats. These are important enhancements of the stakeholder relationship enabled by Web 2.0. However, there is one transformation made possible by digital innovation that can truly be described as revolutionary, and that is XBRL—eXtensible Business Reporting Language—whose use by the largest 500 U.S. companies was mandated by the Securities and Exchange Commission (SEC) beginning in April 2009. A version of XML reporting language, XBRL is based on a system of meta-tagging data, organized through a series of so-called taxonomies that define which tags mean what in terms of a company's financial statement and balance sheet.

XBRL is designed to make a company's financial data more searchable, more comparable, and to be interactive. The new system will benefit not only government agencies but also investors. Prior to the introduction of XBRL, financial analysts who typically develop their own models were required to update these models manually whenever they received new data from one of their investment companies. Even when this information was received via a digital channel, the data itself was not automatically mergable with the spreadsheet being used by the analyst.

XBRL makes any financial data using its tags instantly integratable. It makes voluminous data deeply searchable and queryable. As the investor or analyst works with XBRL data, he or she is able to test different financial scenarios by varying any field in the data table without having to manually recreate any data at all. The alignment of the XBRL taxonomies makes it significantly easier for investors to compare one company or industry sector with another, identify inconsistencies or warning signals in the data, and customize their own early-warning signals by type of data. XBRL also reduces the effort that companies make in producing financial information for different purposes—regulatory filings, their websites, or printed financial statements.

In view of the conservative nature of the investor relations field, it has taken many years to get XBRL off the drawing board. Indeed, some CFOs undoubtedly feared that XBRL would put too powerful a tool in the hands of investors. However, as XBRL becomes the universal standard, we believe that companies will want to work actively with investors to augment these tools in pursuit of greater all-around transparency.

Companies are exploring social media methods such as Facebook and Twitter to communicate with investors while still complying with the SEC's guidelines for insider trading and Regulation Fair Disclosure (Reg FD). (See Guidelines 10 and 11 of this book in Part 5 below for a detailed discussion of Investor Relations planning, sustainability, and transparency. See also "The Dark Side of Web 2.0 and Investor Relations" later in this chapter.) This will be especially true for companies not currently tracked by financial analysts. Since the marginal cost of providing investment analysis for additional companies will drop radically when the manual inputting or transformation of information is eliminated, it will be possible for research houses to offer research on many more companies.

Finally, XBRL helps accomplish a key aim of the SEC: to level the playing field for individual and institutional investors. By making high-quality data available faster for everyone and at less cost, the reliability and value of detailed analysis will be significantly enhanced. For the investor relations professional, the reduction in time required for the mere production of financial information for stakeholders will facilitate a more intimate and strategic relationship, reducing the stress of sudden surprises and enabling longer and more trusting engagements.

Web 2.0 and Investor Relations Ecosystems

Many of the tools previously described will have a powerful impact on the practice of investor relations. However, the biggest challenges faced by the profession today relate to the increased complexity of the IR ecosystem. It is as if each recognized participant in the flow of financial information had grown a dozen heads: print financial media have sprouted video feeds, blogs, and comment fields; investor chat rooms, long a bane of Investor Relations Officers (IROs), have evolved into a maze of interlinked blogs; new media, such as Twitter and its clones Brightkite and Yammer, are showing up with financial communications content. These new micromedia have been embraced by mainstream companies for a variety of communications uses, including investor relations.

In 2009, eBay began tweeting its own analyst calls, answering questions posed on Twitter by investors during the call and augmenting responses delivered live on the conference audio line by the company's CEO and CFO. Here is a brief excerpt from eBay's Twitter feed:

1. I had to go back and update the Griff post…totally missed his "first day at eBay" answer originally. Updated: http://tinyurl.com/c47b6k2:16 PM Feb 25th from web
2. @cianw not from me but I'll find out what I can get my hands on. Hope all is well. Cheers! 1:52 PM Feb 25th from Tweetie in reply to cianw@charleneli already heading back to HQ.…Sorry to miss you #accel12:52 PM Feb 25th from Tweetie in reply to charleneli
3. OK ridiculous. Still no wireless.…I'm heading back to HQ. John was great up there. No slides, no notes. Just conversation. Good stuff. 12:38 PM Feb 25th from Tweetie
4. Someone asked how we address the need to stay connected with our community.… Hello! I'm right over here! No other companies blogging here 12:24 PM Feb 25th from Tweetie
5. This sitting down "fireside chat" format is great.…Need to do this more often http://twitpic.com/10r3212:13 PM Feb 25th from Tweetie
6. Just been informed that wireless will be up in 15 minutes. Murphy and his law can go spit! 12:10 PM Feb 25th from Tweetie
7. JD stressing that there can be multiple winners in ecommerce.…eBay, Amazon, Walmart.com, Zappos.…Love that he included Zappos! :) 12:08 PM Feb 25th from Tweetie

As this series makes clear, much of the tweeting is of a simple and direct kind, but the outlines of a more intimate conversation can be glimpsed in some of the exchanges. Arguably, this kind of banter enhances relationships and helps overcome some of the stiffness inherent in investor relations practice. Time will tell whether tweeting earnings calls are here to stay.

The Dark Side of Web 2.0 and Investor Relations

In the midst of the enthusiasm surrounding the use of new media tools, there is also a dark side that should be of concern to investors and corporations alike. Chief among these must be questions about the permanent accessibility of ad hoc comments by management on new media platforms. Not only could some of this saved "data" become content for shareholder lawsuits; it could also provide another means to compare present performance with past promises. Is the world ready for this permanent record? Attorneys for eBay, at least, aren't so sure. As pointed out on the website Web IR Report, the company is committed to retaining the twitter stream and analyst conference call records for only 90 days. The blogosphere might regard this as a rather shallow retention policy, but one sympathizes with the company's concerns. One thing is certain: As more and more companies create "informal" investor relations content, the question of retention will require a thoughtful policy review.

For many investor relations groups, however, it is the potential for negative rumors promulgated by short sellers that is the biggest concern. These concerns may well have a reasonable foundation.

Employee Communications

In his influential works *Growing Up Digital* and *Grown Up Digital,* Don Tapscott[2] deftly describes the ways in which the Millennial Generation is changing communications in the workplace. Having instant messaged from childhood onward, texting as much or more than phoning, and participated avidly in a variety of social media platforms, this generation would appear poised to deploy the full arsenal of new media tools in the workplace.

Yet corporations have, in large measure, reacted with significant ambivalence to the spread of Internet use of various kinds on the job. In part, as Ron Alsop describes in *The Trophy Kids Grow* Up,[3] this ambivalence is fueled by the skepticism of older generations of managers regarding the utility and productivity of web communications usage. In addition, it is by no means universally accepted that the blending of the personal and professional that emerges as a practical matter in workplace use of social media is a positive development. Our research with Millennials in the workplace[4] suggests that even this supposedly web-obsessed generation displays a broad spectrum of opinion, ranging from those who find the invasion of "work" into social networks abhorrent and others who claim not to understand what all the fuss is about.

On the management side of the equation, the acceptability of social and other new media tools in the workplace is by no means considered a universal good. Even in companies where Internet access is more or less universal, managers have erected significant barriers to accessing social media sites or even prohibit their employees from surfing the web in any form (see Figure 3.2). While almost all companies block pornography on their servers, many others block a wide range of other sites. In recent research by writer Dave Munger and psychologist Dr. Greta Munger of Davidson University, nearly 40 percent of the corporate employees surveyed reported some Internet restrictions at their place of work. More than 20 percent stated that specific sites (e.g., Facebook) were restricted, 18 percent reported "no personal email policies," and 10 percent of respondents were barred from YouTube.

Other research by the Mungers indicates that employers are experimenting with a variety of other types of restrictions, including hours in the work day when the Internet is available, special places (libraries) where Internet access is unlimited, or making web surfing a feature of "casual Fridays."

Much of the corporate concern stems from fears that, given access to the web, employees

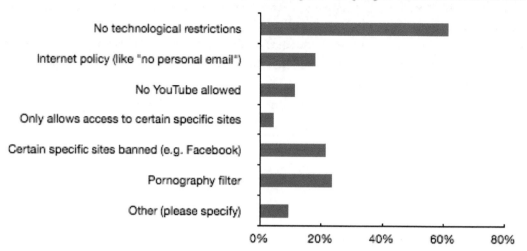

Figure 3.2 How Does Your Employer Restrict Internet Use?
Source: Cognitive Daily, January 23, 2009

will waste time in the workplace shopping online or create liability for the company if illegal or inappropriate blog posting or photo uploading takes place. These fears appear to be somewhat exaggerated. According to the Pew Internet and American Life Project's March–April 2008 survey of 855 employed Internet users (See Figure 3.3)[5] even though three out of four workers have made online purchases at work, only 2 percent say they blog at work, 3 percent admit to watching web video in the office, and none of those surveyed report playing online games during working hours.

Faced with this less-than-universal approbation of social media and other Web 2.0 tools in the workplace, we can turn to the question of whether and how corporations are using any of the new media tools to communicate with their employees. Certainly, given the air of breathless hype with which journalists have embraced social media tools such as Facebook, wikis, blogs, and micro-blogging, we would expect to find plentiful examples of companies using these tools.

And indeed, many examples of the use of these tools can be found. BestBuy, the consumer technology retailer, reports that employees communicate about customer problems and innovation ideas while playing online videogames. The company's employee social networking site, Blue Nation, connects employees on the front line with the customer with headquarters researchers developing new products and new service ideas. Steve Bendt, Senior Manager for Social Technology at the company, reported to a Churchill Club gathering in June 2008 (cited on "The Workplace" blog) that BestBuy employees who used the site had an 8–12 percent turnover rate, far lower than the average for employees as a whole. Oracle's social media platform, Connect, had 8,000 users within 24 hours of launch and by 2008 had more than 10,000 users. IBM claims to have 28,000 employees on "Beehive," but also 15,000 internal wikis. Former Sun Microsystems' CEO, Jonathan Schwartz, is reported to have received 400,000 hits on his blog per month. (More details about the IBM experiment can be found in the case study in Chapter 1.)

Do you ever use the internet to…?			Do you generally do this …			
	have **ever** done this	have **not** done this		at work only	at home only	both at work & at home
buy a product online, such as books, music, toys, or clothing	76%	24%		3%	53%	19%
watch video on a video-sharing site like YouTube or Google Video	53%	47%		3%	37%	12%
send "Instant messages" to someone who's online at the same time	41%	59%		6%	22%	12%
use an online social or professional networking site like MySpace, Facebook, or LinkedIn	35%	65%		3%	25%	7%
read someone else's online journal or blog	33%	66%		3%	22%	8%
play online games	28%	72%		—	24%	3%
contribute writing, files, or other content to your employer's website	23%	77%		10%	3%	8%
create or work on your own online journal or blog	12%	88%		—	9%	2%

Figure 3.3 Online Activities at Work and at Home
Source: Pew Internet & American Life Project Survey, March–April 2008.

These are impressive numbers. And at a handful of companies with a particular orientation toward social media platforms, or an underlying business logic for investing in their growth, we can observe real traction. In the corporate world more broadly, however, the picture is still mixed. The Edelman PeopleMetrics survey of 2006 contains some sobering statistics. Even allowing for growth since then, 26 percent of respondents stated that their companies prohibited instant messaging at work. Only a third of the respondents claimed that their company communicated with employees via blogs, podcasts, and intranets combined. Not a single respondent reported the use of wikis, IMs, chat rooms, or other online forums. While 47 percent reported wikis as useful or very useful for employee communications, as well as 40 and 29 percent for podcasts and blogs respectively, the study showed that in actual practice email communications completely swamped any other channel or volume of usage.

One comparison tells the whole story: While only 16 percent of respondents rated email as the most effective form of communication with employees, 53 percent reported that email was the most common form of employee communication at their companies, twice as many as mentioned their intranets, the next most common form of employee communication.

As an aside, it is worth noting that the survey population accurately gave the highest marks for effective communication to in-person communication, but sadly and predictably, only 12 percent described it as their company's most common type of employee communication.

In the light of these conflicting cross-currents, what are the new media practices that we think will find a permanent place in the arsenal of employee communicators? In order to answer this question, it is necessary to predict the resolution of two separate but equally important issues currently very much under discussion in the workplace:

1. Will the blending of personal and professional communications continue to grow so that communicating with employees on Facebook will be a logical extension of that growth?
2. Will companies continue to create obstacles to workplace access to social networking platforms and other multimedia channels and have policies that discourage individual employee blogging?

There are limits to the extent to which employees feel comfortable sending and receiving information through networks they also use as their personal and social networks. Similarly, the blending of the personal and professional on corporate networks that has grown explosively in the past decade is shrinking as employees spend more and more time communicating electronically in both spheres. It is simply more comfortable to keep these rigorously separated.

However, corporations, mindful of the immense power and adhesiveness of social networks, will replicate the features of these networks in their communications. Rather than use their own pages on Facebook to communicate with employees, companies will create their own internal Facebooks.

There are two primary reasons that it is this style of communication with employees that will come to predominate. The first is that the applications most commonly used in social media such as Facebook perfectly mimic the employee relationship needs of large corporations:

- broad and rapid dissemination of information that can be automatically updated;
- the creation and nourishment of a corporate culture that is highly motivating, promotes productivity, and increases workplace satisfaction;
- the ability to poll employee opinion ad hoc and periodically in an anonymous, aggregated, searchable way;
- a need for forums in which employees can tell stories using all the multimedia tools at their disposal and a means to stimulate innovation and new ideas in a collaborative manner.

The second reason is that employees, having grown very familiar with Facebook-style communications, will embrace these new corporate social media platforms with far more enthusiasm than they have adopted corporate intranets or wikis. If we are correct, then, to butcher Bernard of Chartres, corporations "will see farther because they stand on the shoulders of mice." *Computer mice.*

Community Relations

It may seem counter-intuitive that a virtual global medium should have a significant impact on something concrete and local like community relations. However, the definition of community has changed, and Web 2.0 has proven to be a remarkably powerful tool for engaging with local stakeholders wherever in the world those may be found. In order to do this effectively, companies need to understand the ways in which information flows within local communities have been changed by new media.

Until relatively recently, these information flows were dominated by local news weeklies, a regional metropolitan daily, local community cable programming, and radio stations. In a somewhat broader sense, local community ecosystems were also bound together by sports teams, institutions such as museums and orchestras, local universities, and religious institutions. It is within this ecosystem that corporations played the role of sponsor or grantmaker, supporting the efforts of their own employees to enrich the life of the community.

Today these local news sources are increasingly under threat from community websites and bulletin boards that offer all of the features of a weekly newspaper, as well as a level of immediacy, intimacy, and interactivity that few traditional media can match. And their number is proliferating rapidly. An informal survey of the blogs of New York City, for example, finds that there may be as many as 1,000 bloggers blogging in or largely about Brooklyn. There is a website—www.placeblogger.com—whose principal purpose is to catalogue and aggregate information from blogs all over the world that focus on content about specific places. On April 2, 2009, its home page featured a directory of events in the Kickapoo Valley community of Southwestern Wisconsin and a news item about a science day from www.ventnorblog.com, a blog about a town on Britain's Isle of Wight.

What are these placeblogs really about? www.baristanet.com, the largest placeblog for Montclair, NJ (pop. 37,000) offers a good example. Founded by a former *New York Times* reporter and a partner, Baristanet provides a forum for the discussion of local news and issues, a directory of local activities and events, a place for local businesses to advertise and for citizens of the community to post comments and ask questions. Founded in May 2004, Baristanet is recognized as one of the leading placeblogs in the United States, with more than 5,000 daily visitors. It has inspired placeblogs from Pittsburgh to New Haven.

During any given week, Baristanet will feature local crime reports, obituaries, and restaurant openings—in fact, everything that you would find in a local weekly except that it is interactive and immediate. Here's an item from March 4, 2009:

> Seems like just yesterday we were talking about hellish commutes (oh wait, *it was*). Today's morning commute got off to a rough start with a suspension in service earlier today that put all train service from NJ into NYC to a halt. Service to Penn Station from New Jersey has been restored, but delays are expected.
>
> *6:42 A.M.:* Northeast Corridor, North Jersey Coast Line and Midtown Direct train service has resumed operating into and out of New York Penn Station, however customers may experience residual delays as a result of the earlier service disruption when Amtrak was unable to close Portal Bridge.

The immediacy of this posting is echoed by others about a big factory fire in a neighboring town, alleged racial slurs reported within hours of occurring at a council meeting, and efforts to rally real-time opposition to a proposed construction project. In addition to its own content, Baristanet is a point of integration for relevant web content ranging from New Jersey Transit to Essex County Sports to Boy Scout Troop 12.

By contrast, the website for the local newspaper, *The Montclair Times*, is a conventional repackaging of the print edition online, with daily updates between print editions.

The richness of the interactivity, immediacy, and diversity of placeblogs such as Baristanet poses new challenges but also presents new opportunities for the community relations endeavors of corporations. The placeblogs hold up a giant mirror to the community, offering real-time insights into the needs, concerns, and passions of local citizens. By tapping into these rich conversations, community relations departments of large local employers can develop stronger ties to the community than has been possible in the past. The power to respond to local concerns immediately, whether about an unpleasant odor emanating from a manufacturing facility or a mistaken rumor about a company's job cuts, puts new tools into the hands of communicators and, handled with skill, becomes another source for building corporate reputation.

It is essential that corporate communication professionals understand the rules governing relations with these new media. Are they the same as dealing with a local print newspaper or radio station? In some ways they are, but in others they are subtly different. In creating good relations, the communications professional should always be focused on a win-win for both parties, and this is no different when dealing with new media. However, some of the metrics for achieving this look a little different. Whereas a local newspaper is looking for increased readership through interesting information the communicator might provide or perhaps increased pass-along statistics, the blogger has other metrics that come into play.

Analytics blogger Avinash Kaushik, author of the blog "Occam's Razor," summarizes some of these new metrics as follows:

1. Raw author content (how many new articles)
2. Audience growth (visitors/unique visitors)
3. Off-site growth (how many new RSS/newsfeed subscribers)
4. Conversation rate (comments per post)
5. Citations (Technorati authority ranking)

In other words, creating positive relationships with new local media requires an expanded focus on the metrics that drive them.

Metrics, however, are not the only key difference. Whereas local newspapers or cable stations often think of themselves as the voice of the community, it is actually placeblogs and the conversations taking place on them—painful, irreverent, accusatory, slanderous, and often not instantly sympathetic to an approach from a corporation. The background of the bloggers often dictates the extent to which the blog adheres to conventional journalistic practices, but the wise public relations practitioner should adopt a cautious posture in reaching out to these media. Here are a few key guidelines:

- Expect your motives to be questioned and the nature of the outreach to feature in the coverage.
- Be patient with postings that don't match the courtesy of conventional media.

- Develop a tone that is consistent with the culture of the medium. This will probably be less formal than official press announcements but still needs to be consistent with the corporate brand.
- Assume that local content will find readers *all over the world*, especially in your other plant communities.

Although handling these new local platforms requires learning a few new tricks, the classic rules of communications still apply: know your audience and their interests; offer respect and transparency; and build long-term relationships.

Public Affairs

In no single sector of communications is the potential for the use of social and other new media more palpable than in public affairs. The very definition of grassroots campaigning has been rewritten by the campaign of President Barack Obama, whose new media work, including mybarackobama.com, broke every conceivable record for fundraising and participation. Over the course of the presidential campaign, Obama amassed more than $500 million in small online contributions and accumulated 844,000 MySpace fans. His tweets were followed by more than 118,000 Twitter fanatics. There have been 132 million views of his YouTube channel.

Orchestrated by Chris Hughes, one of the co-founders of Facebook, President Obama's Web 2.0 strategy deployed a full panoply of web tools to enlist, engage, and motivate millions of supporters. Mybarackobama.com was a model of simplicity. Taking its stylistic cue from Facebook, "MyBO" offered its users a way to create their own networks of Obama supporters in their friend base and made it extraordinarily easy for users to sign on for a variety of activities. These included donating money, raising money from others with a personal phone bank, hosting events, attending town hall meetings, or participating in get-out-the-vote drives. Each of these activities was tracked by the software, and users received prompts to help them complete the assignments they had signed up for.

On top of these features, MyBO also connected users with Obama supporters on their other networks such as MySpace and Facebook, as well as serving as a personal repository of text, video/audio, and photography for content about the campaign and the user's interaction with the campaign. In addition to connecting the user with his or her own friends who were supporters, mybarackobama.com also gave them an opportunity to blog to the entire network about their own experiences with the campaign and share thoughts and ideas about the campaign for change.

In the wake of President Obama's successful campaign, congressmen and congresswomen of every political persuasion have taken to the ether with a vengeance. Using tools such as Qik, they are live video-blogging from the floor of the House of Representatives, issuing tweets on an hourly basis, and starting YouTube channels to cement their relationships with their constituencies. While there was some controversy in 2008 (and partisan sniping) about the use of social media by members of Congress, it appears that the use of these tools during the presidential campaign laid to rest fears that these tools would be abused.

In light of the wholesale embrace of a wide range of new media tools by elected officials, it is fitting to ask how these tools will be adopted by corporations and other institutions seeking to influence public policy. Some of the earliest adopters have been grassroots advocacy groups such as Americans for Tax Reform, a right-wing pressure group that opposes tax increases for any purpose. Their website offers links to Facebook pages, Twitter, and RSS feeds. The site itself is completely searchable by state and locality, and all the content is meta-tagged.

These eruptions of citizen protest haven't missed social media tools themselves. When Facebook changed its terms of service (TOS) in February 2009 in an attempt to retain rights to its users' content in perpetuity, even if they closed their accounts, Julius Harper, a Los Angeles-based graphic designer, was able to gather more than 140,000 people to protest the change. Within two weeks, Facebook had agreed to return to its former TOS, pending a review of its policies, and to obtain more feedback from users.

Clearly, Facebook users are more acutely attuned to flash causes of this kind than the average citizen, but it is probably only a matter of time before corporations and other organizations figure out how to use social media to advance their own causes. In the days before ubiquitous broadband access, grassroots campaigns were painstakingly stitched together through direct mail databases, phone banks, and expensive advocacy advertising.

In an ironic twist on one of the best-known examples of public affairs campaigns, the successful resistance by the health care industry to President Bill Clinton's proposed insurance reform in 1993, the "Harry and Louise" characters who appeared in anti-reform commercials then have re-emerged, but this time on the pro-reform side of the equation. Since it is 2009, they are fully garbed in social media clothing with their own website, www.harryandlouise.org and their own fan page on Facebook.

Today, "Harry and Louise" are using social media tools to:

1. Create a campaign theme with a powerful rallying cry and construct a www.harryandlouise.org web community that enables supporters to create their own friend networks to engage the people and organizations they know
2. Enlist these interlocking communities to blog about their views, email, and post to congressional websites
3. Use web analytics to demonstrate the rising volume of public concern
4. Make the web community a jumping-off point not just for web-based activities but face-to-face town meetings and other gatherings
5. Deploy YouTube video interviews to help instruct supporters on how to debate the issues

If the social media lessons of the 2008 presidential campaign prove to be replicable in other public affairs campaigns, there will be one determining factor that, to our minds, will stand out: the propensity of social media channels to give individuals a greater sense of personal ownership of an issue. The ability to activate and motivate their own networks, create their own content in support of an issue, and share their opinions with communities of like-minded individuals should be a powerful force in the public affairs battles to come. This will not be without its challenges for companies and organizations trying to enforce message discipline in pursuit of their legislative or regulatory objectives.

To take just one example, a broad range of organizations and individuals with vastly different ideological perspectives are in favor of legalizing the sale of marijuana. The digital personal content that would be generated in support of such legislation would put serious strains on any coalition in support of this political goal. A cursory search on Facebook of the term "legalize marijuana" brings up 500 individual pages devoted to this subject, ranging from groups interested in medical research to "UPS: United Pot Smokers," which borrows the emblem of a well-known shipping company. However, we believe that as experience in conducting such campaigns grows, the perceived benefits in terms of campaign success will outweigh the challenges.

Web 2.0 and Beyond

Many extravagant claims have been made about the impact of social media and mobile media on our communities. Comparisons to Gutenberg's printing press, the telegraph, and television are claims that overreach. These communications (relationships?) media are in their infancy. Many experiments with them, such as the "twittered" analyst call or corporate "facebooks," may prove to be too costly, wherein the cost/benefit ratio simply fails to support continued employment of such tactics for a particular purpose.

However, some uses of these technologies will prove to be lasting, some perhaps not yet invented. Only time will tell what the enduring innovations will be. From our perspective, this suggests that small investments in exploring even fringe-sounding new media are a valuable expenditure for companies seeking to strengthen stakeholder relationships in a cluttered information marketplace. A healthy skepticism about the use of these tools by corporations is a rational response, but as more young workers who grew up with social media enter the workforce, it will be a bold—perhaps even foolhardy—company that doesn't become deeply familiar with the challenges and the opportunities these media represent.

Notes

1. www.internetworldstats.com/stats/htm, December 2008.
2. Don Tapscott, *Growing Up Digital: The Rise of the Net Generation* (New York: McGraw-Hill, 1998); and *Grown Up Digital: How the Net Generation Is Changing the World* (New York: McGraw-Hill, 2009).
3. Ronald J. Alsop, *The Trophy Kids Grow Up* (New York: Wall Street Journal Books, 2008).
4. Porter Novelli, "Millennials," Summer 2008.
5. Pew Internet and American Life Project, March–April 2008, www.pewinternet.org/Reports/2008/Networked-Workers, 56.

CHAPTER FOUR

Strategic Ethical Relationships

Trust and Integrity

For decades, popular movies have cast corporate leaders in the role of evil, greedy villains. The corporation in such movies as *The International, Michael Clayton,* and the *Quantum of Malice* is a ruthless, win-at-all-costs engine of destruction. And documentaries such as Michael Moore's *Roger and Me* (about General Motors), Alex Gibney's *The Smartest Guys in the Room* (about Enron), and Joel Bakan's *The Corporation: The Pathological Pursuit of Profit and Power* reinforce the popular distrust and skepticism of corporations.

The disparity of power between the corporation and the individual further heightens the lack of trust in business. According to the joint Arthur W. Page Society–Business Roundtable Institute for Corporate Ethics' *Special Report—The Dynamics of Public Trust in Business* (2009):

> "Public trust in business" roughly describes the level and type of vulnerability the public is willing to assume with regard to business relations. Today, a large portion of the public believes that the majority of its vulnerability in business relationships is not voluntary but rather results from a sizable power imbalance that enables executives and companies to assume far less risk than the average person. This sense has been exacerbated by the current financial crisis, in which American taxpayers have been called upon to shore up financial institutions whose risky behavior put the financial system at risk.[1]

In light of popular opinion about business corruption and the perceived power gap between the corporation and the individual, corporations have struggled to re-articulate and re-invent the compact between the individual and the enterprise in a variety of formal and informal ways.

This chapter reviews some of the more significant efforts in this direction set in the context of continuing concerns about business ethics. We discuss the emergence of formal "principles-based" codes of conduct, as well as the ongoing effort to rebuild public trust through the practice of Corporate Social Responsibility. As the examples we discuss show, the strategic adaptations taking place are an attempt to "normalize" the relationships between these social entities by making them both more transparent and more explicit in the context of new regulatory schemes in both foreign and domestic operations.

The Impact of the Scandals of the 21st Century

From the collapse of Enron in the fall of 2001 to the admission of guilt by Bernard Madoff for the largest Ponzi scheme in history in early 2009, trust in business to do the right thing, and in government to enforce laws on the books, has been deeply shaken. Enron's Ken Lay died before he could be sent to prison. Jeffrey Skilling and Dennis Kozlowski are serving time for corporate fraud. Bernard Madoff will most likely never leave a jail cell.

The United States does not have a corner on the scandal market. In January 2009 Satyam (ironically the Sanskrit word for truth) Computer Services' Chairman Ramalinga Raju admitted that he had "cooked" the company's books for years, sending yet more shock waves through a global business community reeling from a financial meltdown that began in 2008.

Why so much fraud by so many executives for so long? It is that we are more aware of it rather than that there is more of it. It also reflects a bit of hubris and the pursuit of more, according to Jim Collins, author of *How the Mighty Fall: And Why Some Companies Never Give In*.[2] Some common and often surprising explanations—not excuses—for the unethical behavior of individual corporate executives include a poor self-image that links a person's worth to salary. There is also the myth that executives have so much responsibility that they think, "I deserve it." Often their fantasies of success go unchecked, and they believe that they have the position of power through society's blessing. Competitiveness can push some executives to unethical actions. Others see themselves as alone in making decisions and therefore seek counsel from no one. Boredom is often another explanation for their actions. Finally, there is the age-old Machiavellian explanation that power corrupts.

Whatever the explanation, the result of these scandals has been deep suspicions about corporate behavior, corporate ethics, and the moral rectitude of CEOs. Corporate credibility has been seriously eroded, and corporate reputation rendered increasingly fragile. The power of NGOs (non-governmental organizations), their influence, and their credibility has grown as a result. And there is a shift in the U.S. Congress, and governments around the world, to increase regulatory action and take a harder line on corporate transgressions.

As the levers of power shift to the hands of legislators and NGOs, corporations need to rethink their response to accusations of fraudulent and unethical behavior that can threaten the corporation's reputation. The time allowed to respond to emerging crises—now minutes rather than days or weeks—is a serious conflict with traditional corporate processes that place a value on deliberation.

Corporate officers increasingly recognize that reputation risk is greater than legal risk, and that lawyers alone can no longer drive an organization's crisis response. The speed of response is essential, as are streamlined processes. The response must address not only "commitment" and "expertise" (traditional), but "transparency" and "empathy" (new).

A corporation's response to accusations needs to demonstrate four critical characteristics of the company:

Commitment. For example, "We are committed to fixing this problem." Or "We are committed to getting to the bottom of this."

Expertise. Statements such as "We have assembled a world-class team of experts." Or "We have hired the global consulting (law, accounting) firm to investigate and report findings."

Transparency. For instance, "We will continue to report to you as new information is obtained."

Empathy. Statements that could express this, such as "We regret the inconvenience/difficulties/upset/pain this has caused."

In addition to addressing new public demands in crisis situations, corporations are increasingly aware that they demonstrate their values through their behaviors. Their actions, not their statements, speak most loudly. Public skepticism represents an opportunity for corporate action. The joint Arthur W. Page Society–Business Roundtable Institute for Corporate Ethics' *Special Report—The Dynamics of Public Trust in Business* (2009), for example, recommends that "leaders must develop a keen practical understanding of the three core dynamics of trust: 1) Mutuality—that is based upon shared values and interests; 2) Balance of Power—where risks and opportunities are shared by parties; 3) Trust Safeguards—that limit vulnerability in the context of power imbalances."[3]

Principles vs. Rules; "Moral Hazard" and the "Animal Spirits"

This need to develop new ethical platforms has generated much debate about the difference between a "rules-based" and a "principles-based" approach to ethical conduct. For corporations and individuals as well, the West Point Honor Code, "A cadet will not lie, cheat, or steal, or tolerate those who do," provides a simple, memorable, and vivid template for ethical behavior using a principles-based approach. Put simply, do the right thing, and when you witness others who do not, do something.

As Compliance Officer Dr. Tim Hedley of KPMG observed,[4] corporations require employees to sign an ethics agreement every time their contract comes up for renewal, usually annually. And they require employees to take ethics training, usually an online session. They are also asked to sign a statement indicating whether or not they have observed unethical behavior on the part of anyone else. This gives most people pause, because if they have seen something, then they must sign and thereby do something about it. If they do not sign, they are lying.

The difference between the principles-based and the rules-based paths are not trivial. Judge Sven Erik Holmes explains that ethics and compliance

are different but complementary. Ethics is to compliance in much the same way principles are to rules. The former are more general, sometimes set at a slightly higher level of aspiration. The latter are more specific. In a business, a notion like integrity, which is an ethical notion, sets a broad standard—a principle to which everyone aspires. But this principle alone won't be effective unless it's backed up by sound rules and procedures. Yet all the rules and procedures in the world are not going to succeed if there are people who are determined to get around them and don't have a deeper sense of integrity.[5]

The reinvigorated focus on such ethical questions stems from the perverse consequences of the development of seamless and highly volatile capital markets in the 20-year period of almost uninterrupted economic expansion beginning in 1987. How is it that fraud came to be so widespread? To a significant degree, corporations and their leaders abandoned a principles-based business model ("do the right thing") in favor of a rules-based one ("if it is not specifically outlawed, then it is OK"). Such thinking created, according to some observers, an environment analogous to the "moral hazard" condition, in which people take greater business and ethical risks because they know that they will not suffer personally from the consequences. It is this "moral hazard" in the syndication of high-risk mortgages that has been held responsible for the global financial meltdown of 2008–2009. This "recklessness" is increasingly seen as the necessary accompaniment to the "animal spirits" of global capitalism.

Robert Shiller comments on the importance of these "Animal Spirits," an idea presented by John Maynard Keynes in his book *The General Theory of Employment, Interest and Money* (1936), written during the Great Depression. Keynes had

> a deeper, more fundamental message about how capitalism worked, if only briefly spelled out. It explained why capitalist economies, left to their own devices, without the balancing of governments, were essentially unstable. And it explained why, for capitalist economies to work well, the government should serve as a counterbalance.
>
> The key to this insight was the role Keynes gave to people's psychological motivations. These are usually ignored by macroeconomists. Keynes called them animal spirits, and he thought they were especially important in determining people's willingness to take risks.[6]

The Great Depression saw widespread bank failures—20 percent of them in the United States— and lawmakers saw the speculation of the banks in the 1920s as a cause of the stock market crash. Bank officials were called before the Senate Banking and Currency Committee to answer for their role in the crash. Then, in 1933, the Glass-Steagall Act was introduced by Senator Carter Glass (D-Va.) and Congressman Henry Steagall (D-Ala.) to limit the conflict of interest created when commercial banks could underwrite stocks or bonds. The law banned commercial banks from underwriting securities, creating separate commercial and investment banks. The law also created the Federal Deposit Insurance Corporation (FDIC) to insure bank deposits and strengthened the Federal Reserve's control over credit.

So devastating were the consequences of the Great Depression that the regulatory framework established under Glass-Steagall endured for more than 50 years. However, as capital markets became more global in the 1980s and 1990s, American financial institutions bridled increasingly under statutes from which their European and Asian competitors were exempt.

Then, in 1997, "after 12 attempts in 25 years, Congress finally repeals Glass-Steagall, rewarding financial companies for more than 20 years and $300 million worth of lobbying efforts. Supporters hail the change as the long-overdue demise of a Depression-era relic."[7]

It is in part the consequences of that repeal that have set the stage for companies to revisit a principles-based approach and think anew about the place of ethics codes in the presentation of their corporate brands.

Ethics Codes

Our institutions have been under stress since well before the recent scandals and the financial meltdown. A crisis of belief had been building. The evidence is apparent in the rising level of doubt in institutions and a cacophony of voices questioning them. Skepticism rising from a shaken belief in the integrity of our institutions appears to be everywhere. Concerned leaders have taken steps to win back the trust placed in them by the public at large and to reverse the quiet crisis that threatens trust—the fragile thread that holds our social fabric together.

In this climate of skepticism, doubt and questioning have quickly replaced belief and trust. The skepticism underpins the perception that the leaders of these institutions have lost their willingness to "do the right thing." The solutions floated for consideration and public debate often focus on the creation of new rules to replace values, seemingly abandoning the notion that self-directed ethical behavior has any potency in the current environment.

This ethical fragility has also caused communications professionals to revisit their own standards of ethical conduct. The Arthur W. Page Society, for example, a professional organization with a single mission—to strengthen the management policy role of the chief corporate public relations officer—has focused fresh attention on the Page Principles:

- Tell the truth.
- Prove it with action.
- Listen to the customer.
- Manage for tomorrow.
- Realize that a company's true character is expressed by its people.
- Conduct public relations as if the whole company depends on it.
- Remain calm, patient, and good-humored.

It is a simple, powerful code of ethical behavior, not just for its members, but for every professional.

A similar re-emphasis on fundamental ethical principles can also be seen within corporations. For example, Johnson & Johnson has used its Credo (See Figure 4.1) to guide its corporate actions and business decisions for more than 60 years. As a company they make the essential connection between the credo, their corporate culture (see Chapter 5 below for a further discussion of corporate culture), and the capabilities of their leaders as the foundation of their ethical decision-making process.

The corporation sees the alignment of its professionals with a values-based process for making important decisions. That alignment process creates common values for the company and its employees, resulting in greater autonomy for employees.

The need to re-assert fundamental values after the scandals at Enron and WorldCom prompted the creation of a new organization. In January 2003 the PR Coalition, an association of more than 20 organizations dedicated to public relations, public affairs, and corporate communication, met in a first-ever summit to determine actions that would meet the challenge of restoring trust. The PR Coalition issued a White Paper that was distributed to the CEOs of the Fortune 500 entitled "Restoring Trust in Business: Models for Action." The three broad actions recommended in the paper—(1) adopt ethical principles, (2) pursue transparency, (3) measure trust—are just as valid in the aftermath of the financial crisis of 2008–2009. The White Paper offers questions to guide companies toward developing and restoring the public's trust in business. (See Figure 4.2)

OUR CREDO

We believe our first responsibility is to the doctors, nurses and patients, to mothers and fathers and all others who use our products and services. In meeting their needs everything we do must be of high quality. We must constantly strive to reduce our costs in order to maintain reasonable prices. Customers' orders must be serviced promptly and accurately. Our suppliers and distributors must have an opportunity to make a fair profit.

We are responsible to our employees, the men and women who work with us throughout the world. Everyone must be considered as an individual. We must respect their dignity and recognize their merit. They must have a sense of security in their jobs. Compensation must be fair and adequate, and working conditions clean, orderly and safe. We must be mindful of ways to help our employees fulfill their family responsibilities. Employees must feel free to make suggestions and complaints. There must be equal opportunity for employment, development and advancement for those qualified. We must provide competent management, and their actions must be just and ethical.

We are responsible to the communities in which we live and work and to the world community as well. We must be good citizens—support good works and charities and bear our fair share of taxes. We must encourage civic improvements and better health and education. We must maintain in good order the property we are privileged to use, protecting the environment and natural resources.

Our final responsibility is to our stockholders. Business must make a sound profit. We must experiment with new ideas. Research must be carried on, innovative programs developed and mistakes paid for. New equipment must be purchased, new facilities provided and new products launched. Reserves must be created to provide for adverse times. When we operate according to these principles, the stockholders should realize a fair return.

Figure 4.1 Johnson & Johnson's Credo Guides Its Actions and Ethical Decision-Making

A SCORECARD FOR RESTORING TRUST

ADOPTING ETHICAL PRINCIPLES

* Has the Board of Directors accepted responsibility for monitoring a corporation's performance in serving all of its key stakeholders—customers, employees, the community and investors?
* Have the CEO and senior management made their corporate values explicit to all employees, making it clear that they are willing to be held accountable for adhering to those values?
* Has the company de-emphasized short-term earnings objectives in favor of metrics that enable investors to better understand the drivers of long-term value creation?
* Have we encouraged and promoted ethical behavior in the company and its operating environment, making it a shared responsibility involving all key constituencies?

PURSUING TRANSPARENCY

* Have we created a process for transparency and disclosure that is appropriate for the company in both current and future operations?

* Have we set our own social and environmental performance targets, defined what transparency means to them and built a case for our approach?
* Have we proactively engaged our stakeholders in dialog about transparency and disclosure?
* Have we monitored our external environment so that we can understand, prioritize, and respond to expectations?
* Have we published our corporate governance policies on our Web site?
* Have we formed an internal committee to ensure that the Board is getting a complete picture of the company's performance?
* Have we established a disclosure committee to evaluate internal controls, review disclosure policies and practices, determine the materiality of information that might need to be disclosed, and review all public communications and SEC filings?
* Do we require employees to take ethics training and sign letters upholding our business principles?
* When addressing issues of public concern, do we localize the message by involving company employees in affected communities and enlisting the help of third parties?
* Are we willing to disclose all of our business, social and political activities, as long as doing so does not raise legal issues or jeopardize our competitive position in the marketplace?
* Do we address the tough questions (e.g., CEO compensation) directly and completely and talk candidly with employees about how and why the company does business the way it does?
* Have we conducted a "culture audit" to ensure that employees believe they are being rewarded for positive behavior?

MEASURING TRUST

* Do we have a program to measure trust that is tailored to the constituencies, objectives and cultures of the company?

Figure 4.2 Guidelines for Restoring Public Trust in Corporations
Source: Guidelines for Restoring Public Trust in Corporations[8]

The Corporation and Its Role in Meeting the Challenge of Rebuilding Trust

The loss of faith in the ethical corporation did not begin with the implosion of Lehman Brothers in September 2008, or with the Enron scandal in the fall of 2001. But it is clear that these events created a crisis of confidence in corporate America and heightened demands by regulators, investors, legislators, and ordinary citizens for meaningful reform of corporate governance, corporate disclosure, and regulatory enforcement and oversight. The notion that markets regulate themselves has been undermined, perhaps even shattered, by a passion for growth and profit that has proved to be unsustainable. As corporations attempted to rebuild the public trust, new instruments emerged to measure their success in this endeavor.

"Edelman Trust Barometer"

One such instrument was the Edelman Trust Barometer, which debuted in 2007 at the World Economic Forum in Davos and quickly established itself as an annual public benchmark.

Matt Harrington, President and CEO at Edelman US, presented the findings of the "2009 Edelman Trust Barometer" at the CCI Symposium on Reputation—Trust Me? in March 2009.[9]

According to the Edelman study, trust in business is country-specific, and it is down to 38 percent in the United States, its lowest level including post-Enron. Trust in every institution declined in 2009 from 2008 in the United States—NGOs by 18 percent; business by 23 percent; media by 15 percent; and government by 13 percent. Trust in every industry declined in the United States—banks (down 35 percent) and automotive (down 30 percent) led the way; technology (down 6 percent) had the smallest decline. Banking and insurance were among the least-trusted industry sectors globally.

Executive inequality, economic crisis, poor business performance, and ethical concerns were the top reasons for the decline in trust in business. Leading issues were the financial credit crisis, energy costs, global warming, and access to affordable health care.

Globally, more people blame government for limiting access to affordable health care, but in the United States, blame is split equally between business and government, according to the study. The United States assigns more responsibility to business than does the rest of the world for solving the problems of energy costs, the financial credit crisis, and access to affordable health care. Business has lost the ability to lead unilaterally, and in the United States, opinion is split, but the majority believes that businesses need to partner with others. By a 3-to-1 margin globally and 2-to-1 in the United States, informed publics agree that government should impose stricter regulations and greater control over business across *all* industries.

Trust in information sources and spokespeople declined across the board. Experts are the most trusted spokespeople for businesses. Expertise and peer conversations drive credibility of information sources. The perceived credibility of all traditional media, digital media, and corporate media as sources of information about a company declined in the United States in 2009. The study observed that to be informed, the public needs information from multiple sources and multiple voices. And it needs to hear the information three to five times to believe it.

The Edelman Study concludes that diminished trust impedes business's license to operate; that regulation does not equal abdication; and that a public company serves shareholders *and* society. A partnership among all interested parties is the solution.

Harrington offered public engagement—an active outreach to multiple groups and stakeholders with an interest in the company—as a strategy for changing policy and communication in four specific areas:

> **Private sector diplomacy.** Companies should actively work to shape policy on the world's most pressing problems, including the ones that don't concern the company's specific industry. They should partner with government and NGOs to drive decision-making and to set strategy on major societal issues.
>
> **Mutual social responsibility.** Companies should embrace actions that benefit both society and the bottom line. They should integrate approaches to societal problems into products and services, as well as involve employees and customers in decisions

and actions about their company's social responsibility.

Shared sacrifice. Corporate leaders should set a collaborative tone for the company in order to emphasize that "We're all in this together." Adopting equitable compensation (executive pay cuts, bonus forfeiture) sends a powerful message. Leaders should actively communicate with and welcome the voices of employees in building this important relationship.

Continuous conversation. Corporate leaders should be agile and "of the moment." They should strive to inform, not to control the conversation. Actions must align with words.

Voices in the Background

Contemporary concerns about the values of the modern corporation may have been stimulated by the post-2008 recession but did not spring out of nowhere. For several decades, the debate about the reconciliation of free market capitalism has been raging both inside and outside the walls of the enterprise. Charles Handy's *The Hungry Spirit: Beyond Capitalism—A Quest for Purpose in the Modern World*[10] is a good example of this debate from inside the system. As an internationally recognized management guru, Handy's book deals with the question of values and personal fulfillment. His discussion is the complex reevaluation of the purpose of business by the son of a British clergyman who pursued a corporate life and that of a management consultant faced with the notion that money is indeed not everything, nor is it merely enough to sustain one's life. His argument set the stage for later descriptions of sustainable business models focused on the triple bottom line—financial, social, and environmental performance.

The importance of trust as an underlying principle that makes for social and economic prosperity was also presented clearly and forcefully by Francis Fukuyama in his *Trust: The Social Virtues and the Creation of Prosperity.*[11] In 2002, a year after the Enron scandal, he noted that to reestablish trust required a balance of both formal and informal trust. Formal trust includes the rule of law, transparency, and publicly evident rules. Informal trust is culturally defined by values and norms that allow people to communicate and deal with others who share those values.[12]

The global importance of the crisis in corporate reporting was articulated in *Building Public Trust: The Future of Corporate Reporting*, by Samuel A. DiPiazza, Jr., CEO of PricewaterhouseCoopers and co-author Robert G. Eccles (2002)[13] and underscored in DiPiazza's *Financial Times* editorial "Good Financial Reasons to Invest in Openness."[14]

Not Just a Corporate Issue

The movement toward greater transparency presents a difficult issue: how to define an effective measure of transparency. It turns out that measuring its opposite—opacity—can be much more fruitful. In 2001, PricewaterhouseCoopers published its "Opacity Index," a year-long study of how 35 nations were affected by their willingness to embrace transparency and openness or their determination to remain opaque. For the purposes of that study, opacity was defined as the absence of clear, accurate, formal, easily discernible, and widely accepted practices in five principal areas:

- *Corruption* in government bureaucracy that allows bribery or favoritism

- *Laws governing contracts or property rights* that are unclear, conflicting or incomplete
- *Economic policies*—fiscal, monetary, and tax-related—that are vague or change unpredictably
- *Accounting standards* that are weak, inconsistent, or not enforced, thus making it difficult to obtain accurate financial data
- *Business regulations* that are unclear, inconsistent, or irregularly applied

The results of the PricewaterhouseCoopers research provided objective, quantitative support for an intuitively obvious conclusion. Opacity has serious economic effects: it *decreases* the quantity and *increases* the costs of capital flows such as foreign direct investment and sovereign bond issuances. In practice, that means the more "opaque" a country, the more it must spend to attract foreign investment and the more it must pay to borrow money.

Countries that embrace transparency will inevitably be more successful at attracting foreign investment. And foreign investment provides not only needed capital for the host country, but, more important, new technology, managerial expertise, and marketing know-how. One could argue that there is a direct link between transparency and a higher standard of living, as well as more and better jobs.

In 2009, the battle for transparency in international capital markets was taken up once again as regulators debated the level of information flows that should apply to offshore financial havens and the appropriate privacy laws governing banking institutions all over the world.

Anti-Bribery, Anti-Corruption, and the Foreign Corrupt Practices Act (FCPA)

Not surprisingly, the effort to combat opacity in sovereign government institutions, of which corruption was a major factor, prompted a parallel effort directed at corporations in the form of the Foreign Corrupt Practices Act (FCPA). In general, the FCPA prohibits corrupt payments to foreign officials for the purpose of obtaining or keeping business. In addition, other statutes such as the mail and wire fraud statutes, 18 U.S.C. § 1341, 1343, and the Travel Act, 18 U.S.C. § 1952, which provides for federal prosecution of violations of state commercial bribery statutes, may also apply to such conduct.

As a result of the Securities and Exchange Commission investigations in the mid-1970s, over 400 U.S. companies admitted making questionable or illegal payments in excess of $300 million to foreign government officials, politicians, and political parties. The abuses ran the gamut from bribery of high foreign officials to secure some type of favorable action by a foreign government to so-called facilitating payments that allegedly were made to ensure that government functionaries discharged certain ministerial or clerical duties. Congress enacted the FCPA to bring a halt to the bribery of foreign officials and to restore public confidence in the integrity of the American business system. Nevertheless, some countries continue to shun global standards and are rife with corrupt business practices.

The FCPA has had an enormous impact on the way American multinational firms do business. Firms that paid bribes to foreign officials have been the subject of criminal and civil enforcement actions, resulting in large fines and suspension and debarment from federal procurement contracting, and their employees and officers have gone to jail. To avoid such consequences, many firms, as part of their ethics policies, have implemented detailed compliance programs intended to prevent or detect any improper payments by employees and agents.

Following the passage of the FCPA, Congress became concerned that American companies were operating at a disadvantage compared to foreign companies who routinely paid bribes and, in some countries, were permitted to deduct the cost of such bribes as business expenses on their taxes. Accordingly, in 1988, Congress directed the executive branch to commence negotiations in the Organization of Economic Cooperation and Development (OECD) to obtain the agreement of the United States' major trading partners to enact legislation similar to the FCPA. In 1997, almost 10 years later, the United States and 33 other countries signed the OECD Convention on Combating Bribery of Foreign Public Officials in International Business Transactions. The United States ratified this convention and enacted implementing legislation in 1998.

The anti-bribery provisions of the FCPA make it unlawful for a U.S. citizen, and certain foreign issuers of securities, to make a corrupt payment to a foreign official for the purpose of obtaining or retaining business for or with, or directing business to, any person. Since 1998, these provisions have also applied to foreign firms and persons who take any act in furtherance of such a corrupt payment while in the United States.

The FCPA also requires companies whose securities are listed in the United States to meet its accounting provisions. (See 15 U.S.C. § 78m.) These accounting provisions, which were designed to operate in tandem with the anti-bribery provisions of the FCPA, require corporations covered by the provisions to make and keep books and records that accurately and fairly reflect the transactions of the corporation and to devise and maintain an adequate system of internal accounting controls.

FCPA ANTIBRIBERY PROVISIONS

The FCPA makes it unlawful to bribe foreign government officials to obtain or retain business. With respect to the basic prohibition, there are five elements that must be present to constitute a violation of the Act:

The FCPA potentially applies to *any* individual, firm, officer, director, employee, or agent of a firm and any stockholder acting on behalf of a firm. Individuals and firms may also be penalized if they order, authorize, or assist someone else in violating the anti-bribery provisions or if they conspire to violate those provisions. Under the FCPA, U.S. jurisdiction over corrupt payments to foreign officials depends upon whether the violator is an "issuer," a "domestic concern," or a foreign national or business. Finally, U.S. parent corporations may be held liable for the acts of foreign subsidiaries where they authorized, directed, or controlled the activity in question, as can U.S. citizens or residents, themselves "domestic concerns," who were employed by or acting on behalf of such foreign-incorporated subsidiaries.

The FCPA prohibits any corrupt payment intended to *influence* any act or decision of a foreign official in his or her official capacity, to induce the official to perform or fail to perform any act in violation of his or her lawful duty, to obtain any improper advantage, or to *induce* a foreign official to use his or her influence improperly to affect or influence any act or decision.

The FCPA prohibits paying, offering, promising to pay (or authorizing to pay or offer) money or anything of value.

The prohibition extends only to corrupt payments to a *foreign official,* a *foreign political party* or *party official,* or any *candidate* for foreign political office. The FCPA applies to payments to *any* public official, regardless of rank or position. The FCPA focuses on the *purpose* of the payment instead of the particular duties of the official receiving the payment, offer, or promise of payment.

The FCPA prohibits payments made in order to assist the firm in *obtaining* or *retaining business* for or with, or *directing business* to, any person.

THIRD-PARTY PAYMENTS

The FCPA prohibits corrupt payments through intermediaries. It is unlawful to make a payment to a third party while knowing that all or a portion of the payment will go directly or indirectly to a foreign official. *The term "knowing" includes conscious disregard and deliberate ignorance.* The elements of an offense are essentially the same as described above, except that in this case the "recipient" is the intermediary who is making the payment to the requisite "foreign official."

Intermediaries may include joint venture partners or agents. To avoid being held liable for corrupt third-party payments, U.S. companies are encouraged to exercise due diligence and to take all necessary precautions to ensure that they have formed a business relationship with reputable and qualified partners and representatives.

Source: usdoj/criminal/fraud/pu:dlj (December 2004 update)

Corporate Citizenship and Corporate Social Responsibility

The various strands of voluntary and regulatory efforts to restore and maintain trust in corporations come together in a renewed emphasis on corporate citizenship and corporate social responsibility. Put simply, Corporate Responsibility (a broader, more strategic term than Corporate Social Responsibility, or CSR) means running a business so that it meets or exceeds the ethical, legal, commercial, and public expectations that the society has of its business organizations. The commercial and legal expectations are clear to everyone, but the ethical and public ones are not so readily apparent. Corporate Responsibility represents to some degree a return to an earlier understanding of the responsibilities of the corporation that was out of favor for two decades.

In 1970 Milton Friedman wrote that "there is one and only one social responsibility of business—to use its resources and engage in activities designed to increase its profits so long as it stays within the rules of the game, which is to say, engages in open and free competition without deception or fraud."

The *New York Times Magazine*, September 13, 1970, featured an article by Milton Friedman entitled "The Social Responsibility of Business Is to Increase its Profits." He declared that "there is one...without deception or fraud."

Such a strong sentiment against any corporate social obligation gave a green light to some corporations to throttle back on their corporate giving activities, as well as their community involvement. As we have seen, the ethical concerns posed by the scandals of the last decade have caused this view to recede.

For several years Corporate Social Responsibility (CSR) has been the term used to describe the non-financial actions of corporations. The Global Compact and SustainAbility defined it this way in 2004: "By [CSR] we mean an approach to business that embodies transparency and ethical behavior, respect for stakeholder groups and commitment to add economic, social and environmental value." CSR means aligning business operations with fundamental and universal values. By its nature CSR is a long-term strategy through which these values become internal to all of the organization's operations, relationships, culture, and identity, transforming them in the process. In the course of incorporating the principles of corporate social responsibility into every facet of the enterprise, corporations have discovered that CSR becomes an essential component of risk management and, as such, a fundamental driver of business and economic success.

A sound approach to CSR helps to build trust in key markets worldwide, identify and address issues before crisis, mitigate risk factors and legal liabilities, contribute to global development, improve employee morale, improve corporate governance, facilitate access to capital, and enhance corporate/brand reputation.

It is certainly now a mainstream expectation that multinational corporations act as socially responsible organizations. On a purely personal level, the obligation of an individual citizen to the larger society is the part of the social contract that makes civilization work. Society grants rights, and individuals have obligations to earn those rights. Corporations as influential members of our society are an integral part of the same social compact.

Beyond the obligation to conduct business legally and make a reasonable profit, many corporations find numerous other positive reasons to go beyond the minimum obligation. Responsible organizations enjoy benefits of their collaborative relationship with the larger society, such as access to capital, increased financial performance, reduced operating costs, enhanced brand image and reputation, increased sales and customer loyalty, increased productivity and quality, increased ability to attract and retain employees, and reduced regulatory oversight.

In practice, corporate responsibility offers several advantages in routine operations of the company. Responsible practices often result in cutting waste, recycling, conserving energy, and the design of products so that they can be reclaimed, refurbished, and reused. Such companies follow a sustainable business strategy. Social initiatives can include partnerships with charities, in which the relationship is good for both the company and the charitable organization. Examples include American Express and its "Charge Against Hunger Campaign" and Home Depot's relationship with Habitat for Humanity, an organization that helps the poor build their own houses with the help of volunteers from the community and businesses.

Internal volunteerism is another example of the benefits of corporations confronting social issues. Organizations that support and actively encourage volunteerism benefit from improved public perception and loyalty, increased employee morale, reduced turnover, and reduced recruitment costs.

Responsible corporations enjoy the benefits to the bottom line that follow from increased customer loyalty. Several organizations in the 1970s and 1980s benefited from taking a pro-environmental position, including Ben & Jerry's ice cream company (now owned by Unilever), The Body Shop, and Tom's of Maine natural toothpaste. These companies, though small, led the way for other organizations to take a positive stand on recycling, products designed to conserve energy, and products that were made from natural and renewable resources.

Workforce issues are also part of the behavior of responsible corporations. Downsizing in the 1970s and 1980s was the management solution to bloated industrial organization. It was the response to double-digit inflation, slumping productivity, and poor product quality. It certainly shook up the workforce, the economy, and the investment community. Subsequent studies demonstrated an unforeseen negative impact on the morale of the surviving workers, evidenced by a survivor syndrome characterized by lower productivity, increased rates of absenteeism, substance abuse, spousal and child abuse, gambling, and hostile and violent behavior at work. In addition, the increases in productivity and profits promised by downsizing did not occur in the vast majority of instances. The community also suffered as downsizing often manifested itself in plant closings, restructured operations, and mergers with other companies, as well as consolidation of operating and manufacturing functions.

Loyalty to the organization was certainly lost in the reshuffling of corporate life. Employees were informed that the company no longer expected to provide employment to its workforce. Schools and universities began teaching their graduates not to expect to work for the same organization throughout their careers. Instead they would be going into a volatile work environment that would make them change jobs every three to five years. They were to consider each appointment as preparation for their next position, most likely at another company.

Customer loyalty was also under pressure. Many products, such as electronics, appliances, and automobiles, began to be very much alike. The advantages of a well-known and respected brand began to evaporate in the face of items that were now marketed as commodities.

Trust in the corporation among consumers suffered in the wake of poor workmanship and mediocre quality. The workforce lost trust in the corporation in the wake of downsizing, mergers, and restructuring. As a result of corporate citizenship efforts, trust in corporations by consumers and employees was making a comeback until the recession of 2008.

Workforce motivation, another casualty of corporate restructuring and downsizing, has benefited from open and honest communication among managers, owners, employees, and investors. Making the relationship between the company and its employees clear through constant and interactive communication has improved employee morale. The workforce has changed considerably as well. It is more and more characterized as a contingent workforce. People work for themselves on a contractual basis to a larger organization. When the job or contract is up, it can be renegotiated if necessary, or the relationship ends.

In practice, corporations create a culture of responsibility with mission, vision, and values statements that align with corporate behavior and cultural values. Their corporate governance policies emphasize ethics and responsibility. Responsible companies assign executive management responsibility to senior corporate officers or a corporate committee. For such companies, corporate responsibility is part of strategic and long-term planning including clear goals and measures for progress. There is also general accountability in that the responsibility goals are part of job descriptions, which makes all employees part of the effort.

Communication, education, and training outline responsible behavior expected of company employees. Employee recognition and rewards help promote responsible behavior. Assessment of performance through a responsibility audit on a regular basis emphasizes its importance. Leaders in corporations use their management positions to model and influence behavior and align their actions with policy as a result. (See Guideline F later in this work for a list of organizations devoted to Corporate Responsibility issues. The mission and contact information are included.)

UN GLOBAL COMPACT

The United Nations Global Compact was launched July 26, 2000, as a truly global corporate citizenship initiative. It is a multi-stakeholder platform. It is voluntary. Its guidelines for companies are rooted in universally accepted principles of human rights, labor, the environment, and anti-corruption derived from:

* The Universal Declaration of Human Rights adopted by the United Nations in 1948
* The International Labor Organization's Declaration on Fundamental Principles and Rights at Work adopted in 1998
* The Rio Declaration on Environment and Development adopted in 1992
* The United Nations Convention Against Corruption adopted in 2000

The Global Compact's mission is to make its Ten Principles an integral part of the strategy, operations, and culture of business. It is intended to facilitate cooperation and dialogue among global stakeholders and to provide a platform for collective action in support of global development objectives.

Human Rights

Principle 1: Businesses should support and respect the protection of internationally proclaimed human rights; and

Principle 2: make sure that they are not complicit in human rights abuses.

Labor Standards

Principle 3: Businesses should uphold the freedom of association and the effective recognition of the right to collective bargaining;

Principle 4: the elimination of all forms of forced and compulsory labor;

Principle 5: the effective abolition of child labor; and

Principle 6: the elimination of discrimination in respect of employment and occupation.

Environment

Principle 7: Businesses should support a precautionary approach to environmental challenges;

Principle 8: undertake initiatives to promote greater environmental responsibility; and

Principle 9: encourage the development and diffusion of environmentally friendly technologies.

Anti-Corruption

Principle 10: Businesses should work against corruption in all its forms, including extortion and bribery.

Figure 4.3 The Global Compact asks companies to embrace, support, and enact, within their sphere of influence, a set of core values in human rights, labor standards, the environment, and anti-corruption.
Source: www.unglobalcompact.org/Portal/Default.asp

The Global Compact can act as a Value Driver for companies to combat poor social and environmental performance by providing a guiding framework. This is particularly important in countries where legal systems do not exist, or are not enforced. Companies can use the Long-Term Performance Model advocated by the Global Compact to counteract an obsession with the short-term performance. To combat a lack of transparency, the Global Compact advocates Communication on Progress reports and also suggests linkage to other initiatives such as the Global Reporting Initiative (GRI). To reverse a lack of stakeholder engagement, they suggest that companies undertake a multi-stakeholder dialogue on key issues and provide a learning platform to build collaboration between corporations and stakeholders.

Companies that adopt a proactive corporate citizenship policy and management strategy can expect enhanced reputation with consumers, the public sector, the media, and financial markets. Such actions create shareholder value at acceptable risk levels and serve as an indicator of responsible excellence and leadership. The positive equity built by good citizenship actions can be a source of future cash flow and profits.

The Corporate Obligation

In the final analysis, changing social and economic forces will continue to dictate the delicate interplay between corporations and their citizen and community stakeholders. There will be periods during which the concern for ethical conduct on the part of corporations is more active than at others. However, we believe that the technological and economic forces that have created a global reputation environment for companies will continue to reinforce the benefits of strong corporate responsibility for the foreseeable future.

Notes

1. *The Dynamics of Public Trust in Business—Special Report*, The Arthur W. Page Society and the Business Roundtable Institute for Corporate Ethics, 2009, 6.
2. Jim Collins, *How the Mighty Fall: And Why Some Companies Never Give In* (New York: HarperCollins, 2009), 20–21.
3. *The Dynamics of Public Trust in Business*, 6.
4. Interview with Dr. Tim Hedley, compliance officer at KPMG, February 2009.
5. *KPMG Ethics and Compliance Report 2008*, 4–5.
6. Robert Schiller, "A Failure to Control the Animal Spirits," *Financial Times*, March 9, 2009, 9.
7. "The Long Demise of Glass-Steagall: A Chronology Tracing the Life of the Glass-Steagall Act from Its Passage in 1933 to Its Death Throes in the 1990s, and How Citigroup's Sandy Weill Dealt the Coup de Grace," www.pbs.org/wgbh/pages/frontline/shows/wallstreet/weill/demise.html.
8. Guidelines for Restoring Public Trust in Corporations, www.corporatecomm.org/pdf/PRCoalitionPaper _9_11Final.pdf.
9. Matt Harrington, "2009 Edelman Trust Barometer," CCI Symposium on Reputation—Trust Me?, March 2009, www.corporatecomm.org/archive.
10. Charles Handy, *The Hungry Spirit: Beyond Capitalism—A Quest for Purpose in the Modern World* (London: Arrow Books, 1998).
11. Francis Fukuyama, *Trust: The Social Virtues and the Creation of Prosperity* (New York: Free Press, 1995).
12. "Restoring Trust in a Cynical American Public—Francis Fukuyama," in *Arthur W. Page Society Journal, 2002 Annual Conference: Earning Trust: Aligning Communications and Leadership Behavior*, ed. Edwin Nieder (New York: Arthur W. Page Society, 2002), 6.
13. Samuel A. DiPiazza, Jr., and Robert G. Eccles, *Building Public Trust: The Future of Corporate Reporting* (New York: John Wiley & Sons, 2002).
14. Samuel A. DiPiazza, Jr., editorial, "Good Financial Reasons to Invest in Openness," *Financial Times*, February 11, 2005, 13.

CHAPTER FIVE

Corporate Culture's Increased Signifigance

The emergence of the global networked enterprise in which the "knowledge worker" becomes the core corporate asset has propelled corporate culture into a key role in motivating and directing the activities of the corporate workforce. To a greater degree than ever before, the successful enterprise is based on a culture that is consistent, clear, and constantly reinforced.

More than two decades ago, Peter Drucker predicted the changes that would create the new management approaches and the need for the networked organization:

> The typical large business 20 years from now will have fewer than half the levels of management of its counterpart today, and no more than a third the managers. In its structure, and in its management problems and concerns, it will bear little resemblance to the typical manufacturing company, circa 1950, which our textbooks still consider the norm. Instead it is far more likely to resemble organizations that neither the practicing manager nor the management scholar pays much attention to today: the hospital, the university, the symphony orchestra. For like them, the typical business will be knowledge-based, an organization composed largely of specialists who direct and discipline their own performance through organized feedback from colleagues, customers, and headquarters. For this reason, it will be what I call an information-based organization.
>
> Businesses, especially large ones, have little choice but to become information-based. Demographics, for one, demands the shift. The center of gravity in employment is moving fast from manual and clerical workers to knowledge workers who resist the command-and-control model that business took from the military 100 years ago. Economics dictates change, especially the need for large businesses to innovate and to be entrepreneurs. But above all, information technology demands the shift.[1]

The information-networked organization, along with global markets and access to digital technologies, has clearly made Drucker's prediction a reality for multinational corporations. These forces have certainly had a powerful and transformational impact on the practice of corporate communication.

These changes, notes the World Bank, have been a double-edged sword, with both positive benefits and negative consequences.

> ...the growing integration of economies and societies around the world...has been one of the most hotly-debated topics in international economics over the past few years. Rapid growth and poverty reduction in China, India, and other countries that were poor 20 years ago, has been a positive aspect of globalization. But globalization has also generated significant international opposition over concerns that it has increased inequality and environmental degradation.[2]

Thomas Friedman, in his *The Lexus and the Olive Tree* (1999), called globalization the "inexorable integration of markets, nation-states, and technologies to a degree never witnessed before—in a way that is enabling individuals, corporations and nation-states to reach around the world farther, faster, deeper and cheaper than ever before...the spread of free-market capitalism to virtually every country in the world."[3]

The recession of 2008–2009 called into question the widely taught and accepted belief in self-correcting free markets that goes back to Adam Smith's "invisible hand' in his *Wealth of Nations*. Paul Krugman, in his *The Return of Depression Economics and the Crisis of 2008* (2009), notes that: "financial globalization has definitely turned out to be even more dangerous than we realized."[4] And George Soros observed that "Globalization did not bring about the level playing field that free markets were supposed to provide according to the market fundamentalist doctrine."[5]

Unlike the corporations of the 19th and 20th centuries, which were set up to attract capital and to create manufacturing capability, present-day corporations do not need a manufacturing operation or a way to raise capital. The 21st-century corporation, just as Drucker predicted, harnesses and focuses on human capital. It is people with smart ideas, who know how to innovate and run things.

Corporations are about people, and they are a network of people. For example, Microsoft is powerful as an individual corporation, but it is really as powerful as it is because of its alliances, most notably with Intel, but also with everybody else—large clans of companies. Even in a downturn, a benefit of the global market is that there is much work to be done, and several companies willing and eager to share the work. On any given day, a company buys goods and services from its competition, and sells to its competition. That was the 20th-century aerospace industry's model. Even though that industry had separate companies, they were partners in almost every operation. When Astronaut Buzz Aldrin took his "one small step for man; one giant leap for mankind" more than four decades ago, the accomplishment was truly a public-private partnership of the highest order. The Japanese Keiretsu and the Korean Chaebol are examples of such partnerships.

Because people and their governments had believed for the most part that multinational corporations brought prosperity, an almost unlimited ability to merge, coalesce, divide, and move operations had been given to the corporation. In retrospect, the bargain looks increasingly one-sided, now that there appears to be a crack in the economic progression. The financial crisis provided a realistic picture of the cyclic nature of growth and prosperity. A downturn or recession often follows an enormous boom, similar to the Panic of 1870, the panic of the turn of the last century, the Great Depression of the 1930s, and the recessions of the 1950s, 1970s and 1990s—

10- to 30-year cycles. As baby-boomers begin to retire, a large portion of the population will have left the work force, at least in the United States, Europe, and Japan, underscoring a 20- to 25-year generational cycle.

Contemporary corporate executives may not like everything their corporations do, but by and large they have a job, and therefore they are socialized to believe that the company is basically doing the right thing. Successful corporations realize that as they communicate internally, that if they are talking about endless prosperity, somewhere in their internal or external constituency people are not going to believe them. People sense intuitively that nothing grows forever.

For corporate communication executives, the central functions of the corporation remain, but they are in the hands of a greatly reduced central staff—much like an orchestra—surrounding the CEO. The management "span of control" numbers are greatly increased. The challenges, mitigated by the ubiquity of Web 2.0 intranet and Internet communication, remain:

- Creating a unified culture and vision for the corporation that can be shared by increasing numbers of specialists, complicated by the globalization forces over the last three decades that have negated the contract between the individual and the corporation—much as the "job for life" has disappeared for two generations of corporate employees.

- Motivating the corporation through career opportunities, rewards, and recognition; these traditional human resources (HR) functions now seem more difficult when employees owe no loyalty to the corporation, but their effort is critical to the organization's success.

- Creating a structure that can get work done with a mobile workforce spread over many time zones. The creation of managers who understand that the new communication environment has long since abandoned the "command and control" model of internal communication for the "inform and influence" model and the "be informed and be influenced" model that can work with new forms of social media and Web 2.0.

What is the key to success in organizations that have sustained performance over a long period? The economy? Luck? Corporate excellence? Most executives look at the short term—stock price and quarterly results—and such a focus is understandable, given the speed of economic activity and the basis of the reward system for executives. The investments of hedge funds and the short tenure of CEOs feed the intense focus on short-term thinking at the expense of a long-term strategy.

In an environment of forces that buffet corporations constantly, the sustainable companies are resilient. Such corporations set goals and meet them through professional performance and execution of strategies, supported by a workforce that understands its common vision and shared values. They align their employees with a common purpose. They understand that to accomplish this, they must reinvent themselves as the environment demands and recruit and retain people who are resilient and drive for renewal. They must create and sustain viable corporate cultures, as well as robust processes and metrics to monitor their success, and take corrective actions when needed.

The Culture of the Corporation

Corporations and organizations shape and influence the behavior of individuals in subtle yet powerful ways. These forces, like the wind and the tides in natural environments, are often unseen and unnoticed themselves, but their effects can easily be observed. These forces combine to create the culture of a corporation.

Human groups, in an anthropologist's terms, by their nature have a culture—a system of values and beliefs shaped by the experiences of life, historical tradition, social or class position, political events, ethnicity, and religious forces. In this context, a corporation is no exception. Its culture can be described, understood, nurtured, and coaxed in new directions. But rarely can a corporate culture be created, planned, or managed in the same way that a company creates a product or service. Analyzing the culture of a corporation, when appropriately done, offers powerful insights into the organization's beliefs, values, and behavior.

Terrence Deal and Allen Kennedy popularized the term "corporate culture" in 1982 with the publication of their book *Corporate Cultures: The Rites and Rituals of Corporate Life*.[6] In analyzing a corporation's culture, they provided the descriptive terms:

- artifacts and patterns of behavior that can be observed, but whose meaning is not readily apparent;
- values and beliefs that require an even greater level of awareness;
- basic assumptions about human activity, human nature, and human relationships, as well as assumptions about time, space, and reality.

The assumptions that form the foundation for a corporate culture are often intuitive, invisible, or just below the level of awareness.

Nevertheless, J. Steven Ott offered some "Methodological Approaches for Studying Organizational Cultures" in his *The Organizational Culture Perspective* (1989)[7] that identify three levels of corporate culture. The first, Level 1A, is artifacts, or the physical things you can apprehend with the five senses—things you can see, touch, hear, smell, and taste. Focused observation is generally the fundamental tool for analyzing corporate artifacts by wandering around looking at physical settings, rummaging through archives and other records, looking at organization charts, listening to the corporation's special language, myths, stories, sagas, and legends about the company. Also in the first level is Level 1B, patterns of behavior—for example, how people greet one another, how they conduct meetings, how they travel, play, orient new members, discipline one another, eliminate members who do not fit in, and select leaders.

Examples of artifacts and behaviors, the first level, abound: corporate logos, the company headquarters buildings and "campus," annual reports, company awards dinners, the annual golf outing, the business attire at the main office. The artifacts and behaviors can be observed. Often these are outward manifestations of what the corporation believes and values, no matter what it says its values and beliefs are.

A deeper level, Level 2, is beliefs and values. This level is not to be confused with the written mission, vision, or values statements of a corporation. Those written documents are considered artifacts of the corporation. The values and beliefs of a company can be identified by the alignment of what the company says it believes and values and its actions. For example, Johnson & Johnson (see the discussion in Chapter 4 above) aligns its actions with its Credo. Enron, in contrast, behaved differently than its written statement of values would dictate.

Examples of the corporation's values and beliefs may be articulated in a slogan or an ad campaign, such as Volvo's "For Life." These are simple yet effective ways to put into words a concept that may often be very complex and difficult to articulate. Such slogans present a complex pledge from the company to its customers to create products that improve their lives. Companies that actually write a values statement find the task difficult because the written presentation too often sounds like the values statement of almost any company. Clichés and platitudes can make even the most honest presentation seem hollow.

Level 3, basic underlying assumptions—that is, what informs the actions and beliefs of the company—is even more difficult to articulate because it requires the analysis of both what the company says and an observation of what it does, and then a synthesis to determine conflicting areas. One example of a fatal conflict between the projected basic assumption and what lies beneath the surface is the demise of Enron and Arthur Andersen. Both companies quickly lost their customers' trust when scandals surfaced around their business practices.

Key to identifying the underpinnings of Levels 1 and 2 are iterative interviews with major insiders by an objective person not part of the company. The identity can be revealed using ethnographic and other analytic methods. These may be as simple as asking a key company executive to "tell me a story" or to ask "how do you do that?" Inspection and analysis of company statements and archives can also reveal the basic assumptions that energize the corporate culture.

In diagnosing a corporation's culture, one must study the physical setting, read what the company says about its culture, and test how the company greets strangers. In interviews with company people, the questioner should ask about the company history, ask why the company is successful, why it grows, and who gets ahead in the corporation and why. Interviewees should describe an average day at work and detail how he or she spends their time. Other organizations in the industry should be benchmarked. The interviewer should seek to understand the career path progression and determine how long people stay in jobs, particularly middle management jobs, paying close attention to what people in the company discuss, as well as to the anecdotes and stories that pass through the cultural network.

Of course there can be a dark side to corporate culture. A culture in trouble exhibits symptoms such as: no clear values or beliefs, many beliefs with no agreement on which ones are most important, destructive and disruptive heroes (leaders), disorganized rituals, fragmentation, an inward and short-term focus, morale problems, emotional outbursts, and ingrown and exclusive subcultures that clash.

How can one identify corporate cultures? Deal and Kennedy described four categories as a way to identify and to understand the various types of corporations. They called them Corporate Tribes:

- tough guy/macho culture
- work hard/play hard culture
- bet-your-company culture
- the process culture

Figure 5.1 provides descriptive information about each of the four corporate tribes—examples, risk, feedback from the environment, rewards, people, organizational structure, and behavior.

Culture Type/ Characteristics	tough guy / macho	work hard / play hard	bet-your-company culture	the process culture
Examples	Advertising, Construction, Entertainment, Publishing, Venture Capital	Consumer sales; Retail stores	Oil, Aerospace, Capital-goods, Mining, Investment Banking, Computer Design, Architectural firms; Actuarial Insurance	Banks, Insurance, Financial Services, Government, Utilities, Heavily regulated industries (pharmaceuticals)
Risk & Stakes	HIGH	LOW	HIGH	LOW
Feedback from environment	QUICK	FAST (You get the order or you don't.)	SLOW (Years with constant pressure)	VERY SLOW TO NONE
Rewards	Short-term focus; speed, not endurance	Short-term focus; endurance, not speed	High stakes; constant pressure; long-term focus	Focus on how work is done; Real world remote
People	"COWBOYS"; Individuals; Rule-breakers	Super salespeople are the heroes	Company over individual; heroes land the big one	Young managers seek a "Rabbi" Achieving rank; V.P.'s are Heroes (or survivors)
Structure of Organization	FLAT for fast decision-making	FLAT for fast-decisions; forgiving of poor decisions	HIERARCHICAL; Slow decision-making	HIERARCHICAL (many layers of management); slow decision-making from the top down
Behavior	Informal; Temperamental behavior tolerated; Stars	TEAM Players; informal atmosphere; friendly, optimistic, humor encouraged; NO PRIMA DONNAS	Formal, polite; TEAM players; NO PRIMA DONNAS	Protect the system; "Cover your ass" mentality; Emphasis on procedures, predictability, punctuality, orderliness

Figure 5.1 Characteristics of Four Corporate Tribes
Source: Goodman, Corporate Communications for Executives, 1998[8]

Figure 5.2 compares the dimensions of risk and feedback among the four corporate tribes. Careful observation and analysis of the physical setting of the company, what it says about itself, how members of the company greet strangers, how people spend their time in the organization, the career paths, the length of time people stay in jobs at the organization, the company stories and anecdotes, the jokes people tell, and what people write about or discuss reveals the characteristics of the corporate tribe that will help in understanding the corporate culture.

	Fast	Feedback	
Low Risk	Work hard/ play hard	Tough guy/macho	High Risk
Low Stakes	Process	Bet-the-company	High Stakes

Figure 5.2 Corporate Culture Types Compared by Risk/Stakes and Feedback
Source: Goodman, Work with Anyone Anywhere, 2006[9]

Models for Developing Sustainable Corporate Cultures

In creating a sustainable corporate culture that is adaptable, multinational corporations look at relationships with employees, local affiliates, thought leaders, partners, and communities.

Employees

The goal of positive employee relations is to promote employee retention, increase productivity and work quality, stimulate innovation, and help attract high-quality future employees. In spite of the importance of these goals in increasing shareholder value, employee relations is often an under-appreciated and under-resourced aspect of organizational communications. One reason for this is that relations with and communications to employees reside in so many different places in an organization and are concerned with so many different discrete topics. These include human resources, operations management, training and development, plant management, security, and environment and health, to name a few. In addition, especially in a large-scale organization, it can be very difficult to communicate efficiently so that the entire organization receives information directly from the source (as opposed to the grapevine, which universally operates more effectively than any official medium). In order to confront these challenges, high-quality employee relations depends on four core principles: integration, sequencing, content and medium, and feedback. (For further discussion see Guideline D: Employee Relations in Part 5 below.)

Affiliates

In the increasingly prevalent "network" model of the enterprise, how companies handle their global affiliates has become increasingly important from a culture perspective. In a Pilot Study of Affiliate Relations, CCI Corporate Communication International asked 12 selected multinational corporations to respond to a six-question online survey. Seven corporations answered the questions. Two corporations in the sample had a specific management function dedicated to Affiliate Relations, or a similar role. One commented, "We do have a colleague who is responsible for international communications, and is first point contact for our affiliates. They come to her for any advice they need about corporate and product communications in particular." Corporate communication managers consider it a priority to communicate with local affiliates. They consider the need for affiliate relations to be increasing, because international business is now growing faster, and will increase in the future. One corporate communication executive underscored the growing importance by saying that the "need for consistency in messaging and globalization of the communications function becomes clearer. This topic is also higher on the agenda of the Board. Importance is understood well nowadays." Companies that do not have a dedicated position to manage affiliate relations do this for communication about specific functions and internal communications through key stakeholders within local affiliates, and through regular meetings and message distribution. Some use an extensive feedback loop with local affiliates to ensure quick answers to questions, and others use lots of centralized control, require alignment with corporate messaging, and review all broad messaging in the businesses and regions. To measure communication performance between headquarters and affiliates, corporate communication executives use periodic surveys or occasional audits of local and external messaging for affiliates. It is common for executives to manage relations with local affiliates through meetings (either face-to-face or video conferencing) on a regular basis, weekly update meetings for all communication colleagues by phone, or through regular/daily engagement. (For further discussion see Guideline L: Affiliate Relations in Part 5 below.)

Thought Leaders

In today's environment of rapid technological and economic change, the chief executive is also required to be the visionary for the company, explaining not just where the company is going, but where the world is going. This role has come to be called "thought leadership." At the same time, the primacy of the chief executive's position in a company, analogous to President Teddy Roosevelt's calling the White House a bully pulpit, means that every contact a chief executive makes is an opportunity to enhance the brand and strengthen relationships. In many companies, this opportunity is managed proactively and is called "executive relationship management (ERM)." Thought leadership and ERM, taken together, represent the effective use of the chief executive role to enhance the reputation of an organization with a wide range of stakeholders. It is particularly important in nourishing corporate culture by signaling to employees what matters to the leadership.

The technology revolution of the 1990s brought with it a critical shift in what companies—and, by extension, chief executives—needed to be knowledgeable about. Prior to this period, it was reasonable to expect that senior corporate officers would be deeply intimate with their own companies and well informed about the history and evolution of the industry in which their companies operated. When called upon to make a speech, the chief executive would speak conventionally about the challenges and achievements of his own organization with perhaps a brief rhetorical flourish in the direction of broader issues. At the same time, the public began making new demands on the corporation in the form of calls for greater engagement in what came to be known as corporate citizenship. Spurred by the consequences of rapid globalization, publics in the advanced economies wanted to know where corporate management stood on issues of diversity, human rights, fair trade, the environment, and energy use.

The interweaving of these two strands created an environment in which one of the key attributes of the successful CEO is the ability to offer the company's perspective on a variety of world issues. A well-managed thought leadership program is designed to deliver this perspective as effectively as possible. (For further discussion see Guideline M: Thought Leadership and Executive Relationship Management in Part 5 below.)

Partners

The phrase "transaction communication" refers to a written, oral, electronic, or other communication relating to a proposed merger transaction that is transmitted to a large audience, whether internal or external. It is not intended to cover daily communication among co-workers. During times of uncertainty, it is especially important to communicate with employees to provide current information and to help staff to stay focused on their role in achieving objectives. It also can be important to communicate externally with key business partners, opinion leaders, and consumers. Care must be taken to follow normal leading practices, as well as to account for some special considerations that apply in the context of transaction communication. Corporate affairs officers, with assistance from the professionals in Integration Planning, Human Resources, and legal departments, develop guidelines to assist top leaders and communication professionals in their efforts. (For further discussion see Guideline K: Transaction Communication in Part 5 below.)

Communities

Corporate citizenship initiatives reflect the beliefs and values of the enterprise, and those actions demonstrate them to the company's business and local communities. Such engagement has resulted in a change whereby multinational corporations are much more responsive to communities. Many have adopted sustainability as a business philosophy and long-term strategy to create shareholder value. Companies such as GE, with its Ecomagination initiatives, embrace opportunities and manage risks deriving from economic, environmental, and social developments. They focus on the potential for sustainable products and services, reducing or avoiding costs and risks that come with previous social and environmental practices. The companies that are competent in meeting global and industry challenges make good long-term investments. They remain competitive and enhance the reputation of their brands by integrating sustainability into their business strategy. They meet shareholder expectations for appropriate returns on investment, growth, and transparency. They use resources—financial, social, and environmental—to reinforce customer relationships through innovative products and services. They adopt ethical practices and invest in the development of their workforce. (For further discussion see Guideline F: Corporate Citizenship in Part 5 below.)

Communities around the Globe

To create a corporate culture of accountability, successful global enterprises align operations with fundamental and universal values, such as the ones in the UN Global Compact (see the ethics discussion in Chapter 4 above) on human rights, labor rights, environmental standards, and anti-corruption and transparency. They embody transparency and ethical behavior. They recognize that "soft" issues for business—environment, diversity, human rights—are now "hard": hard to ignore, hard to manage, and very hard to control if they go wrong. They also understand that the intangibles—corporate reputation, governance, innovation, research and development, environmental and social performance, brand equity, human capital, leadership and strategy, product and service quality—drive the corporation's value.

The consequences of not acting as a "public diplomat" exacerbates the intangible risks: poor social and environmental performance; an obsession with short-term financial performance; lack of transparency in corporate reporting; lack of stakeholder engagement; limited risk management of critical issues. This can put the company's reputation at risk, making it the subject of lawsuits (domestic/foreign), negative media coverage, NGO (non-governmental organization) pressure, consumer boycotts, eroding public trust, negative analyst assessment, and market punishment.

Recognizing such risks for U.S.-based multinational corporations, the PR Coalition (see the link to its site at www.corporatecomm.org) in 2007 met and issued a White Paper on Public

Diplomacy, offering models for action for business professionals as public diplomats. These actions (See Figure 5.3) are in four main areas:

(1) Engage in Initiatives That Encourage Mutual Respect & Cooperation
(2) Ensure Employee Attitudes & Behavior Reflect Company Culture & Values
(3) Build on More Interactive Relationships with Counterparts Abroad
(4) Help Foreign Employees & Customers Understand Company Culture & Values.

(1) Engage in Initiatives That Encourage Mutual Respect & Cooperation	(2) Ensure Employee Attitudes & Behavior Reflect Your Culture & Values	(3) Build on More Interactive Relationships with Counterparts Abroad	(4) Help Foreign Employees & Customers Understand Your Culture & Values
✓ Demonstrate a desire to be "local" companies when operating abroad. ✓ Learn more about local ways of doing things, particularly those that provide opportunities for mutual growth and success. ✓ Show respect for all people. Work to overcome divisions & create a social & economic environment where everyone is respected, regardless of race, ethnicity, gender, religion, sexual preference or economic status. ✓ Create partnerships that are welcoming to people of other cultures & that encourage & respect diverse views. ✓ Show how traditions–religious, social and intellectual–have evolved from countries & cultures around the globe. ✓ Create opportunities in other countries through local hires, vendor relationships & economic assistance. Stress education as a ladder to opportunity. ✓ Promote company generosity in support of charities & international causes. ✓ Make public diplomacy actions a corporate officer's responsibility.	✓ Provide professional development programs for company employees stationed or traveling abroad for extended periods. This includes language education, cultural rules and biases, global communication and negotiation styles and conflict styles and resolution. ✓ Create "circles of influence" from relationships with organizations, chambers of commerce, journalists, local business leaders, and expatriates. ✓ Develop global travel practices for shorter-term assignments that include preparation, language training, listening skills and cultural understanding. ✓ Familiarize employees with differences in global business practices. This involves transparency, accounting, management, negotiations and meetings. ✓ Incorporate workforce capabilities in hiring and vendor engagements, reflecting global mindset, languages, cultural intelligence and global business practices.	✓ Hire interns from abroad. ✓ Engage the local business community through active participation in its organizations. ✓ Provide incentives for the non-home country work force to visit on business, as tourists or as students, and for the home country workforce to visit other areas of the world in which the company operates. ✓ Sponsor international short-term assignments for high potential employees. Create competitive scholarships for foreign employees & their dependents to study in the home country. ✓ Participate in existing or create new student exchange programs for employees & dependents. ✓ Become part of the local community through employee volunteerism, strategic philanthropy and greater engagement with NGOs. ✓ Provide financial support for government educational and cultural exchanges (English language training, American Corners, artists on tour) that have proven effective but are currently underfunded. ✓ Identify, contact and develop relationships with alumni of business, educational and cultural exchange programs. ✓ Support the creation of a corps of foreign service officers made up of academics and business people with specialized expertise who could work abroad in short term assignments.	✓ Create employee information programs that provide information that reinforces America's commitment to strong cultural values. ✓ Celebrate national days in countries where the company operates by providing information about each country on websites and work locations. ✓ On national holidays (July 4th for US Corps.), provide information to foreign employees and customers about home country society and culture. ✓ Make company executives more accessible to foreign journalists to help provide more favorable impressions of the home country. ✓ Create opportunities for engagement in foreign countries that will foster a sense of common interests and common values between the home country and people of different countries.

Figure 5.3 The PR Coalition White Paper on Public Diplomacy Offers Models for Action
Source: PR Coalition White Paper: "Private Sector Summit on Public Diplomacy: Models for Action"[10]

To ensure a positive corporate culture, corporate communication executives demand high ethical standards of all members of the corporation. Corporate communication leaders exude believability and communicate clearly. They act to retain, motivate, and inspire employees to meet the challenges of a hostile business environment around the globe. Finally, communication executives assume personal responsibility and accountability for their actions simply because everyone inside the company, and out, expects a corporation and its leaders to do so.

Notes

1. Peter Drucker, "The Coming of the New Organization," *Harvard Business Review* (January–February 1988): 45.
2. www.worldbank.org/economicpolicy/globalization.
3. Thomas Friedman, *The Lexus and the Olive Tree: Understanding Globalization* (New York: Farrar, Straus, and Giroux, 1999), 7–8.
4. Paul Krugman, *The Return of Depression Economics and the Crisis of 2008* (New York: Norton, 2009), 190.
5. George Soros, *The New Paradigm for Financial Markets: The Credit Crisis of 2008 and What It Means* (New York: Public Affairs, 2008), 95. See also his editorial, "Changing the hypothesis: why 'adaptive' trumps 'efficient'" *Financial Times*, November 27, 2009, 7.
6. Terrence Deal and Allan Kennedy, *Corporate Cultures: The Rites and Rituals of Corporate Life* (New York: Perseus Books, 2000).
7. J. Steven Ott, *The Organizational Culture Perspective* (Pacific Grove, CA: Brooks/Cole Publishing, 1989).
8. Michael B. Goodman, *Corporate Communications for Executives* (Albany, NY: SUNY Press, 1998), 32.
9. Michael B. Goodman, *Work with Anyone Anywhere: A Guide to Global Business* (Belmont, CA: PPI, 2006), 149.
10. www.corporatecomm.org.

CHAPTER SIX

Economic Factors

Notwithstanding the current challenges to the world economy, the forces of globalization that have been active since 1945 have had a profound impact on the context of corporate communication. These impacts extend all the way from operational questions, such as the changing nature of specific communications partners (whom we talk to), to subtle cultural issues relating to the changing world order. While issues such as the economic organization of world markets might seem to fall outside the realm of corporate communication, these issues do in fact play a significant role in how organizations manage their communication, where these communications emanate from, and the tone that they adopt.

Global Economic Organization

These issues began, in many ways, at the most basic level: the changing lexicon of world affairs since the end of World War II and more recently the end of the Cold War. After the reconstruction of Europe was undertaken in the late 1940s and early 1950s, the economic and political evolution of the world was largely driven by two forces: the struggle between philosophical systems and the fight against colonialism. The first of these forces pitted the free market economies of the so-called West and, after the Communist victory in China, the Marxist economies of the so-called East—the Soviet Union, the People's Republic of China, and the satellite socialist states of Eastern Europe—against each other. Within this framework, it was natural for "Western" statesmen and commentators to talk of three "worlds": the First World of Western democracies; the Second World of the Communist "dictatorships of the people"; and the Third World, those economies not necessarily affiliated with the other blocs but generally less prosperous on a per capita basis than they.

This glossary, determined largely by political/philosophical orientation, was augmented by the language of economic development that came to be used as the post-colonial nation states of continental Africa and Asia sought to lift their populations out of subsistence levels of economic strength. The terminology of economic development grew out of the efforts by global financial institutions such as the World Bank and the International Monetary Fund to provide aid and assistance. Initially, countries requiring assistance were simply described as "underdeveloped." As distinctions between the economic achievements of different "underdeveloped" countries made it necessary to adopt a more nuanced vocabulary, other terms such as "developing," "less developed" and "more developed" were added. During the 1970s, the term "Newly Industrializing Countries" came to stand for those developing countries that were doing better than the rest of the Third World. In order to avoid politically sensitive distinctions, and as the Cold War rhetoric diminished, other broad terms such as "North" and "South" came into use. In 1981, Antoine van Agtmael, director of the International Finance Corporation's Capital Markets Department, coined the phrase "emerging economies" to counteract what he believed to be an under-active investment climate in Thailand caused by the negative connotations of the term "Third World." In 2001, Goldman Sachs, seeking a term to describe a group of nations the bank believed would surpass the United States, Western Europe, and Japan in economic size by 2050, simply called them the BRIC countries: Brazil, Russia, India, and China.

We cite these evolutions in terminology not because we are here concerned with their accuracy or, indeed, whether one country or another falls into this or that group. We have described the changes in the lexicon because they symbolize for us the underlying attitudes toward different cultures and different countries in a business context. This is important because how we communicate and whom we feel the need to communicate with are deeply rooted in the preconceptions and prejudices expressed in the terminology. The long and largely successful economic journey from the post-World War II global settlement has ingrained in us certain attitudes and communications styles that are increasingly at odds with the reality of the world economy in the 21st century and the power relationships it entails.

It is difficult to grasp today, but with the world in ruins in 1945, the United States accounted for between 40 and 50 percent of the entire world economy. Both the continental allies and the defeated nations of the Axis had suffered profound destruction of their economic capacity. The continental allies had an immediate post-war GDP that was less than 80 percent of its 1939 levels, lower than in the early 1920s. German industrial production in 1946 expressed as output per person was less than one-third of its size in 1936. In fact, the destructive impact of the global conflict had knocked Italy and Japan back to the levels of 1910, Germany to 1890, and Austria to 1870. Collectively the defeated powers had lost between 30 and 75 years of economic development. Even by 1950, after reconstruction was well established, per capita GDP in Europe was the equivalent of that of the United States in 1905. In contrast to the shrinkage suffered in Europe, the U.S. economy grew by 50 percent between 1939 and 1946.[1]

It is not only the economic statistics that demonstrate the massive disparity between the United States and the rest of the world after World War II. After pursuing the wartime lend-lease program instituted essentially to give military materiel to help Britain prosecute the war effort, the U.S. followed up post-war with the Marshall Plan and other related reconstruction programs that provided $14 billion in aid.[2] In the context of these unequal relationships, it is not surprising that the institutions established in connection with the Bretton Woods program—the International Monetary Fund and the World Bank—were dominated completely by the United States.

Amply capitalized and fueled by the consumer products boom in the United States, American multinationals became the engines of world economic growth. Whereas international investment flows in the period before the Great Depression had largely involved investing in foreign government securities and related infrastructure projects, the post-war flows were dominated by foreign direct investment by U.S. corporations. During the 1960s, 30 to 50 percent of the industrial capacity of Latin America was attributable to foreign investment, predominantly American. During the same period, American companies were dominant leaders in the European automotive industry and provided 80 percent of Europe's information technology infrastructure.[3] Foreign direct investment in Europe by American companies rose from $2 billion in 1950 to $41 billion in 1973.

In this context, it is hardly surprising that the behavior and communications styles of corporate America should have shared many features with the "ugly American" tourist stereotype of the post-war period. The failure to adapt to local cultural sensitivities, the insistence that American management and business practices be adopted regardless of the country, and the strong-arm tactics illustrated in horrific detail by the behavior of companies such as ITT in Latin America, all stemmed in different ways from the economic organization of the post-war world.

In response to this dominance, the world's economies experimented with a variety of development models that, in turn, had an impact on the culture and communications environments of multinationals. In order to protect their own industrial champions, many countries, particularly in Europe, established significant tariff barriers to keep American and other foreign goods out. This made it necessary for multinationals to establish local subsidiaries as country-based organizations with their own legal structures and cultures. In some countries such as India, foreign ownership restrictions meant that multinationals had to accept minority positions in exchange for the right to do business in those countries. The combination of economic and legal regulatory factors meant that for much of the post-war period, companies were organized internationally in single-country or regional structures, and their communications flowed from these decentralized structures.

This period of U.S. and then Western European economic hegemony, the Bretton Woods period, lasted until President Richard Nixon uncoupled the dollar from the gold standard in 1971. A small number of countries, notably Japan, had also made astounding strides in this era. At the end of World War II, Japan's economy was 100 years behind the United States. By 1973 it was a mere 10 years behind. In the early 1960s, South Korea was one of the world's poorest countries. By 1996 it had gone from an economy smaller than Ghana or the Congo to surpassing Portugal and Greece. Similar explosions of growth took place in Singapore, Hong Kong, and Taiwan. In the late 1970s, China consciously rejoined the global economy. It returned farmland to private farmers, unshackled industry from government direction, and set up export promotion zones. The country's economic output grew 400 percent in 20 years, and by 1992 China had replaced Japan as the world's second-largest economy. The Indian economy took a similar, if slower, path, growing by 50 percent between 1950 and 1970. In 2005, Goldman Sachs published a report suggesting that India's GDP would surpass that of France by 2020.

What made this growth possible was massive foreign investment, an orientation to building export, and, in the later phases, the transformation of global production. The global economy of the 1990s is the story of how companies from all over the world decoupled their sourcing, manufacturing, and eventually significant parts of their research and development from their country of national origin. In every industry from steel to toys, raw materials, final production, and even customer service moved to those parts of the world able to deliver results at a competitive price: in the case of China, toys, commodity chemicals, and textiles; in the case of Mexico, computer motherboards; in India, customer support and IT services. Nor was this a purely Asian or Latin American phenomenon. Ireland benefited by offering a highly educated workforce of native English speakers to the information technology industry. By the year 2000, Ireland had surpassed the United States as the world's leading exporter of software.[4] Dublin became one of Europe's most prosperous cities.

Impact on Communication Needs and Infrastructure

Other commentators will debate the next phase in this process of economic growth and globalization, which faced a significant reversal in 2009. For our purposes, however, it is important to note that this global economic transformation has had a profound impact on all of the stakeholders with whom the modern multinational corporation has to deal. Whereas it was once necessary to focus public affairs efforts primarily on Washington, D.C., and Brussels, it is now clear that government relations efforts in Beijing, Brasilia, and New Delhi have become equally, if not more, important. The significance of these changes has not been lost on the leadership of multinational corporations that have been building their corporate public affairs resources in China as fast as the available talent will permit. Companies that were once content to run small regional monitoring operations covering the whole of Southeast Asia from Singapore or Hong Kong have established public affairs bureaus in Beijing. The companies with established Beijing communication offices are similarly expanding their public affairs presence in other major Chinese commercial centers such as Shanghai, Guangzhou, and Shen-Chen, revealing a more sophisticated understanding of the power of regional governments in the new Chinese economy.

For companies sourcing the majority of their raw materials and semi-finished goods from Asia, the quality of the supply chain depends on increased transparency and accountability, as the problems with lead paint in Chinese toys and melatonin in pet food have shown. While this topic might appear to be an issue more for quality-control experts than communicators, the consequences for corporate reputation make it clear that it needs to be squarely in the sights of communications professionals. The business process evolution that has taken place in the supply chain thus creates the need for a new kind of vigilance in the reputation risk management efforts of multinational companies. The Timberland Corporation, for example, an enterprise with consumers who are exceptionally alert to sustainability issues, has established an intense review and licensing process that extends to the sourcing and production of every raw material that goes into its products. Working with environmental experts, the company has committed to eliminating the use of PVC in its products where feasible, and to converting to water-based glues to reduce its dependence on solvents. In 2006, according to the company's website, environmentally superior adhesives were used in 8 million pairs of shoes and boots, saving more than 160 tons of solvents. Ensuring that all of its outsourced suppliers adhere to these commitments is a crucial element of the company's brand promise.[5]

This issue of integrity in product manufacturing is not solely the concern of companies with a specific reputation for sustainability to protect. Leading brands of U.S. pet food suffered a severe attack on their reputations when it was discovered that more than 50 brands of dog food and 40 brands of cat food were adulterated with melamine, a toxic byproduct of plastic manufacturing. The melamine had been introduced into animal food processed in China by independent contractors who used it to pass inspection by mimicking protein levels produced from more expensive animal and vegetable sources.[6] In 2009, Lennar, one of the biggest American home builders, became the victim of drywall manufactured by a Chinese subsidiary of Knauf, a German building supplies company. The drywall was determined to be contaminated with sulphur gases that corrode copper in electrical and plumbing components, as well as being a potential health hazard.[7]

Gaining a clear picture of raw material usage and product ingredients has entered the communicator's job description in a new way. Another ingredient under increasing scrutiny is the human labor component. Swiss banking giant Credit Suisse was appalled to discover that thousands of soccer balls it had purchased to give away to youth clubs ahead of Euro 2008, the soccer championship, had been produced by child labor in Pakistan. The company donated $1 million to UNICEF in response to its embarrassment.[8] Clearly, some of these issues were produced by criminal acts that were hard to foresee, but in an increasingly outsourced global economy, product integrity has taken on a whole new meaning.

Similarly, when a large percentage of the manufactured content in a company's products comes from independent factories in Latin America or Asia, employee communication becomes a very different but equally pressing item on the communicator's agenda. This is also true for companies that outsourced their customer service and IT operations to developing countries during the rapid economic expansion of the 1990s. Communicating with workers who are not direct employees poses special challenges. If in addition to those challenges the workers are multilingual and have a culturally distinct relationship to communications technologies, the complexities are multiplied. Since employee communications channels are often rudimentary at outsourced service providers and contract manufacturers, it is crucial to establish clear expectations about how one's company's workflow issues will be communicated to workers. Fortunately, after the Asian business boom of the past decade, managers are becoming more aware of the importance of employee communications. In a June 2009 study by Aon Consulting, researchers discovered that 68 percent of respondents in Hong Kong expected to devote more resources to employee communications. Employee communications was also the number-one priority for respondents in China and Taiwan.[9]

Another crucial economic factor is that with the enormous increase in corporate equity held by sovereign wealth funds and other investors outside Europe and North America, investor relations requires an entirely new set of communications skills. Sovereign Wealth Funds (SWFs), investment entities for national assets, grew exponentially through the commodities boom (particularly energy) of the last decade. Formerly a little-known backwater of the global investment scene, the funds exploded into public view with the 2007 investment by the Abu Dhabi Investment Authority of $7.6 billion in Citigroup.[10] This was followed by $3 billion invested by the China Investment Corporation (CIC) in the New York-based private equity firm Blackstone Group. According to Morgan Stanley, sovereign wealth funds collectively controlled $3.2 trillion in 2008. Even with the collapse in global equity values caused by the financial crisis, Monitor Group estimated in August 2009 that the funds were still worth $1.2 trillion collectively. The largest of these funds are found in the United Arab Emirates, Singapore, Norway, Kuwait, Russia, and China.

The emergence of these huge investment pools out of the transformation of the global economy poses new challenges in investor relations. While the best of these funds, such as the Singapore SWF, Temasek, and the Norwegian fund, are highly transparent about their assets and investment strategies, many of the other largest funds are still somewhat opaque. The funds are not generally activist shareholders in the traditional sense, but a number of them are highly sensitive to the issues of corporate social responsibility. In 2006, for example, the Norwegian fund dropped Walmart from its approved list of investments and sold $400 million in Walmart shares because the fund was dissatisfied with the human rights and labor practices of Walmart and its suppliers.[12] In 2009, the fund delisted the Donfeng Motor Corporation for selling trucks to Burma and placed Siemens under observation because of the bribery scandal involving the company. At the time, the fund owned approximately 1.34 percent of Siemens's voting shares, according to the Norwegian Ministry of Finance.

The other significant concern for investor relations professionals is the widespread fear that SWFs are susceptible to domestic political influence in making and divesting themselves of foreign investments. For the investor relations person, this potential source of investor volatility by a major investor will require a deeper and more sustained understanding of geopolitical forces at work than has historically been the case.

Climate Change, Energy, and Trade Politics

The changing economic scene also has profound implications on the practice of public affairs. Notwithstanding the economic downturn of 2008–2009 and the reduction in the demand for core commodities such as energy and metals, the rapid growth of the BRIC economies has placed new pressures on the world's natural resources. These pressures have emerged just as concerns about climate change and the energy options required to address that issue have reached a new peak of intensity. There are fundamental questions still to be answered about the cost of creating a new energy infrastructure to deal with global warming. The various estimates of the cost are all in the $100-billion-per annum range, and while these figures are daunting, the political realities in finding a consensus on which path to take seem even more insuperable. As the United Nations struggles to create such a consensus, growing economies such as India and China, according to *The New York Times*, are refusing to sign on to any program until the developed world agrees to reduce emissions to 40 percent of 1990 levels by 2020.

Even companies not directly involved in the energy business or in heavy manufacturing will need to take a fresh look at how energy and trade policies have become intertwined. In August 2009, the Brazilian government threatened to break U.S. pharmaceutical patents in retaliation for what it considered unfair American subsidies of domestic cotton, soybean, and ethanol production, all significant Brazilian agricultural products.[13] This example of the politicization of trade and energy issues will undoubtedly have a ripple effect through many different industries as governments seek to satisfy domestic opinion while at the same time pursuing viable geopolitical strategies.

This nexus of energy, food, and trade policies will be more intense than it has ever been in a global economy increasingly characterized by economic equals rather than dominant and submissive parties. This development will, in turn, place new pressures on private multinational corporations to develop and maintain a corporate positioning that can satisfy the needs of this powerful and engaged group of sovereign stakeholders.

We can, perhaps, best summarize the impact of economic factors in the transformation of global corporate communication as follows:

1. What happens outside the "home" country is more important than what happens inside.
2. When a company's major shareholders, supply chain, employees, and customers are in the former Third World, cultural assumptions about communications need to be challenged from top to bottom.
3. The company's corporate communications resource infrastructure needs to be stood on its head—80 percent devoted to global, 20 percent to the home market.
4. What were once discrete and manageable public affairs issues from country to country have melded into a single global polity and need to be managed by the communication discipline from this perspective.

For a discussion of the specific strategies for coping with these transformations, see A Corporate Communication Workshop in the Appendix of this volume.

Notes

1. Jeffrey A. Frieden, *Global Capitalism: Its Fall and Rise in the Twentieth Century* (New York: W.W. Norton, 2006), 261.
2. Ibid., 268.
3. Ibid., 295.
4. Ibid., 420.
5. "Timberland Environmental Stewardship," Timberland Corporation Website, http://timberlandonline.co.uk/chemicals/environ_chemicals,default,pg.html, accessed October 20, 2009.
6. David, Barboza and Alexei Barrionuevo, "Filler in Animal Feed Is Open Secret in China," *The New York Times*, April 30, 2007, late ed.
7. Leslie Wayne, "Thousands of Homeowners Cite Drywall for Ills," *The New York Times*, October 7, 2009, late ed.
8. "Red Faces at Credit Suisse over Child Labor Soccer Balls," Soccernews.com, April 22, 2008, n.p., accessed October 20, 2009.
9. Richard Payne, "Responding to the Crisis," *Aon Consulting* (May 2009). Summary of Survey Findings at http://www.aon.com/attachments/Responding_to_the_crisis.pdf, accessed October 20, 2009.
10. Eric Dash and Andrew Sorkin, "Citigroup Sells Abu Dhabi Fund $7.5 Billion Stake," *The New York Times*, November 27, 2007, late ed.

11. "China to Invest \$3 Billion with Blackstone Group," *Wall Street Journal Market Watch*, May 20, 2007, http://www.marketwatch.com/story/china-to-invest-3-billion-with-blackstone-group, accessed October 20, 2009.

12. Bill Baue, "Norwegian Government Pension Fund Dumps Walmart and Freeport on Ethical Exclusions," *Social Funds Sustainability Investing News*, June 16, 2006, http://www.socialfunds.com/news/article.cgi/2034.html, accessed October 20, 2009.

13. Bradley S. Klapper, "WTO Sanctions U.S. over Cotton Subsidies," *Washington Post*, September 1, 2009.

PART THREE

Managing Public Issues

Models for Corporate Communication Practice

As described in Part 2, the three forces of globalization, technology transformation, and changes in enterprise structure have had a profound influence on the communication challenges faced by global organizations. The new challenges have arisen from the manner in which the three forces have created new power relationships, new stakeholders, and new pathways for communicating with these stakeholders. Among the biggest challenges faced by these organizations (and itself a by-product of these sifting stakeholder relationships) has been the issue of trust and integrity. Ironically, in an era in which state socialism, as a viable competitor to free market capitalism, withered away (to paraphrase Engels), a series of business scandals has significantly reduced the levels of trust between international business and its stakeholders. In response, the global business community and the associations engaged with it have invested much effort in redefining the expectations for trust and integrity. Part 2 explored the various manifestations of the new "compact" between business and society promoting accountability and transparency in the global economy.

As the issues of trust and integrity have moved increasingly into focus, it has also become clear that the primary drivers of accountability and transparency were the intangible qualities of corporate behavior collectively understood as "corporate culture." Corporate culture was not a new concept, and indeed the idea that "the way we do things here" has a vital significance in the life of the firm has deep roots in 19th-century industrial management. However, two factors that increased in intensity in 20th-century enterprise made corporate culture and—above all—an ethical corporate culture an increasingly important dimension of corporate behavior.

The first of these factors is a combination of scale, complexity, and the creation of the global supply chain. To use a simple analogy, when the customer is present to see the merchant's hand pressing down on the scale as he measures out the dry goods, there is a powerful incentive (in the form of the risk of detection) for good corporate behavior. Conversely, when the ultimate customer is thousands of miles away and is buying an intangible product, detecting individual bad behavior becomes orders of magnitude more difficult, and the risk to the "bad actor" is measurably reduced. Add to this the complexity of the modern supply chain, in which tens of thousands of individuals have a hand in creating a product or service. In this scenario, apportioning blame becomes almost impossible.

The second of these factors is the emergence in the world economy of knowledge-based services such as insurance, entertainment, and software, for which the promise of performance relies on a wide variety of factors open to misinterpretation. While it is relatively easy for a buyer and seller to agree on what constitutes a functioning toaster or a cut of fresh meat, satisfaction with knowledge-based services often requires a highly complex set of understandings between the parties. Thus, the ability to generate the trust required to facilitate transactions of this kind demands an even greater reliance on the integrity provided by a high-functioning corporate culture.

Finally, we looked at the specific economic factors that have produced a global ecosystem fundamentally reshaping the relationships among nation-based business organizations, sovereign states, and global non-governmental institutions. This review of the impact of the three forces provides a crucial platform for the examination of the ways in which the corporate communication discipline is being reshaped to meet the emerging needs of the new millennium. However, as a prelude to discussing both strategic and tactical models for corporate communication practice, it seems appropriate to review briefly the historical precedents out of which contemporary practices have grown. This is especially useful because the growth and development of both business management and communications practices have not been simply linear. Between the 1880s, when corporate communication began to emerge as a distinct discipline, through to the present, there have been a series of cycles in which different features of the relationship between the enterprise and its stakeholders have been ascendant. Not surprisingly, these have often coincided with pro- or anti-business eras, but by reviewing the history we can more easily discern the major innovations in practice that have occurred.

The brief historical overview thus enables a detailed view of the principal strategic communications models that exist today, as well as examining how these models are exercised in the day-to-day practice of key communication tactics directed at employees, shareholders, communities, and regulators. Part 3 concludes with a chapter on the critical issue of measurement, which consists of a review of both outputs and outcomes—that is, outputs in the form of the most important metrics for the practice of corporate communication and the resources being invested in the discipline by today's corporations, and also in the form of outcomes measurement, in which some emerging methodologies for measuring reputation over time are discussed.

CHAPTER SEVEN

Precedent

The History of Communication in Corporations

Historians of the practice of public relations are tempted to reach far back in time to document the emergence of the discipline out of political advocacy and propaganda going back to ancient times. They cite Cicero: "I have created a hole through which my client has slipped" and describe in loving detail the "Joyeuses Entrées" of the 16th-century Burgundian dukes. These triumphal processions through the towns of the Low Countries do indeed meet the definition of public relations, designed as they were to bring closer together the duke as ruler and the citizenry as subjects.

However, we will omit a discussion of these remote genealogies to focus on the era that can truly be said to have witnessed the emergence of public relations as we understand it today, an era shaped by the muckrakers Lincoln Steffens, Upton Sinclair, and Ida Tarbell; by Ivy Ledbetter Lee, counsel to John D. Rockefeller; by George Creel and the "Committee on Public Information" (CPI's "Four-Minute Men"); by Edward Bernays and Doris Fleischman; by Walter Lippmann, author of *Public Opinion*; by John W. Hill and Don Knowlton (founders of Hill and Knowlton); by Paul Garett of General Motors; and by Arthur W. Page at AT&T. The foundations for contemporary corporate communication practice were laid in the early decades of the 20th century. Figure 7.1 lists key events of this period. Figure 7.2 links 12 contemporary functions of the corporate chief communication officer with these historical precedents.

Corporate communication or corporate public relations, for our investigation of precedent, can best be defined as the conscious attempt by corporations and other organizations to improve the public's opinion of them in order to assure their freedom to operate as independently as possible. The extent of these freedoms has varied widely, but we can legitimately date the consciousness required to believe in the importance of good public opinion to the era of the great railway barons of the second half of the 19th century. William Henry Vanderbilt's famous October 8, 1882, retort to a reporter that "the public be damned" is too often quoted out of context. He actually said, "The public be damned. I am working for my stockholders."[2]

Year	Key Event
1919	Edward Bernays and Doris Fleischman open their agency Carlton and George Ketchum start Ketchum Publicity
1920	Ivy Lee and Associates formed First commercial radio station broadcasts from Pittsburgh
1922	Walter Lippmann writes *Public Opinion* American Association of Engineers holds "First National Conference on Public Information" *Reader's Digest* begins publication
1923	Claude Hopkins writes *Scientific Advertising* Edward Bernays writes *Crystallizing Public Opinion* *Time* magazine begins publication Oswald Garrison Villard writes *Some Newspapers and Newspapermen* American Society of Newspaper Editors adopts first nationwide Code of Ethics
1924	Advertising Federation of America formed J.H. Long writes *Public Relations: A Handbook of Publicity*
1925	Ivy Lee writes *Publicity: Some of the Things It Is and Is Not* National Better Business Bureau is formed
1926	Roger William Riis and Charles W. Bonner write *Publicity* Society of Professional Journalists adopts first national Code of Ethics
1927	Arthur W. Page works at AT&T as Vice President of Public Relations John Hill open his office in Cleveland Hayes Robins writes *Human Relations in Railroading* Edward Bernays defines "Counsel on Public Relations" in an *Editor and Publisher* ad Harold Lasswell writes *Propaganda Techniques of World War One* Walter Lippmann writes *The Phantom Public* John Dewey writes *The Public and Its Problems*
1928	Edward Bernays writes *Propaganda*

Figure 7.1 The Foundations for Contemporary Corporate Communication Were Laid in the 1920s[1]

It is to this very question—whether a corporation might have a responsibility or duty to individuals and groups beyond its owners—that the emergence of public relations as a consciously practiced discipline provides the answer. As we examine the evolution of corporate communication in the age of the semantic web, it is useful to reflect on the most important milestones of the discipline through the last 100 years of public relations. For while the communication tools available have changed radically, the business and social issues affecting the relations between corporations and their stakeholders have remained strikingly constant.

For much of the latter part of the 19th century, as the great steel, chemical, and railroad trusts were being assembled in the United States after the Civil War, the idea that corporations might need the good opinion of the general public would have seemed laughable. Federal and state action, or inaction, was bought and paid for directly by corporate interests. In his epic history of the period, *Age of Betrayal: The Triumph of Money in America, 1865–1900* (2007), Jack Beatty summons up a descriptive image of Tom Scott, the president of the Pennsylvania Railroad in the 1870s: "Congressmen rustled in Scott's garments, swept up in the trailing cloak of his designs." In describing one of Scott's campaigns to rescue a planned subsidy for the Texas and Pacific Railroad, one of his opponents, Collis Huntington, wrote to one of his operatives: "They offered one member of Congress $1,000 cash down, $5,000 when the bill passed and $10,000 of the bonds when they got them if they would vote for the bill." Beatty's subtitle "The Triumph of Money in America" brilliantly captures this chapter in U.S. history.[4]

Corporate Communication Function	Precedent
advocate or "engineer of public opinion," in support of the company's policies	Creators of Public Relations: Edward L. Bernays (1892–1995), author of *Crystallizing Public Opinion* (1923), *Propaganda* (1928), "The Engineering of Consent" (1947) Walter Lippmann, author of *Public Opinion* (1922), *The Phantom Public: A Sequel to Public Opinion* (1927)
steward of company brand and brand perceptions	Global companies use the company name as the worldwide brand
champion of corporate citizenship (philanthropy) actions	The corporation as good citizen in the locations it is in; the corporation's social responsibility
counsel to the CEO and to the company	The corporation's chief advisor on communication policy, strategy, and actions worldwide; Ivy Lee, Arthur Page
driver of company publicity	Ivy Ledbetter Lee, author of *The Press Today: How the News Reaches the Public* (1929)
manager of company's reputation	Reputation management; *Fortune*'s Most Admired Companies; *Management Today*'s Most Admired Companies; *Financial Times*
manager of company's image	Advertising as a means of projecting company image
manager of employee relations (internal communication)	Recognition that focus on employee's attitude toward the company is key to successful change and fundamental to relationships with constituencies external to the company; Jack Welch and GE
manager of relationships between the company and ALL of its key constituencies	The Corporation as Citizen, Arthur Page and AT&T
manager of relationships between the company and its key NON-CUSTOMER constituencies	Government and community relations, lobbyists; muckrakers such as Ida Tarbell and Upton Sinclair focused public attention on business practices
source of public information about the company	Early corporate practice focused on the CPI Committee for Public Information's four-minute men—local opinion leaders—professionals, lawyers, businessmen, organized to maintain the support of World War I on the home front
support for marketing and sales	Use of advertising and sales as the prime focus of a company, the public relations function acts as support for campaigns

Figure 7.2 Precedents for Corporate Communication Functions[3]

As the 19th century drew to a close, however, the voices raised against what President Theodore Roosevelt was soon to call "malefactors of great wealth" began to be heard more strongly. In 1894, after years of watching and writing about John D. Rockefeller, one of the early so-called "muck-rakers," John Demarest Lloyd, published a book-length study of Standard Oil. On the eve of publication, according to Ron Chernow's *Titan: The Life of John D. Rockefeller, Sr.* (1998), Lloyd felt that the public was ready for his revelations: "The sky seems full of signs that the time for the appearance of such information has come."[5]

Equally persistent was Lloyd's muckraking colleague, Ida Tarbell, whose multipart series in *McClure's Magazine* on the behavior of the industrial trusts won her early renown. When it became known that Tarbell was working on a major exposé of Standard Oil, Lloyd gave her all of his notes and records from a career of observing the company. Her series ultimately ran from November 1902 through August 1905, a tour de force of invective that even President Roosevelt was reading with admiration. Also in *McClure's* in 1906, Ray Stannard Baker's "Railroads on Trial: How Railroads Make Public Opinion" chronicled the process of scanning and clipping newspaper articles and then sending agents to personally visit each editor, noting the editor's political beliefs and views on numerous topics, as well as the focus of the newspaper.

Consistent with the practices and beliefs of the time, Rockefeller and Standard Oil resolutely refused to engage or controvert Tarbell's allegations, in spite of the fact that, as subsequent research has shown, her work contained a number of significant errors. Although the popular press was filled with accounts of Rockefeller's fury and torment at the series, neither Rockefeller nor Standard Oil ever addressed Tarbell's charges.

The popularity of the Tarbell series and other works by opponents of the trusts such as Upton Sinclair derived from many different sources. The later years of the 19th century and the first decade of the 20th saw an upsurge in a middle class readership hungry for "morally uplifting" gossip about the behavior of millionaire industrialists. The growth in the power of organized labor and liberal agitation about working conditions in the great industrial conglomerates all contributed to a vibrant stew of opposition to the interests of the capitalists. The increasing public concern about the power of large corporations led some early innovators to consider a new approach to the rights of the public.

Nowhere was this effort more thoughtfully pursued than at the American Telephone and Telegraph Corporation, whose management understood the importance of preserving the company's right to create and manage a legitimate monopoly. As early as 1903, faced with public perceptions that the "Bell System was a predator seeking to bring down local telephone companies," the company retained the Boston-based Publicity Bureau to place positive articles about the company in newspapers around the country. This program was one of the first examples of a corporation seeking friendly relationships with reporters and editors in order to promote favorable public opinion. Under the leadership of Theodore Newton Vail, who took up the reins at AT&T in 1907, the company committed itself to educating the public about issues of mutual interest in its attempt to dispel the "class prejudice" and "class feeling" that was making Americans increasingly hostile to corporate interests. In seeking tools to create positive "public relations," executives such as Vail took a decisive step away from the antagonistic style of most contemporary executives, who refused to acknowledge any public interest beyond that of the shareholders.

In the United States and in Europe, the years leading up to World War I saw the gradual emergence of labor laws and conventions that would determine the role and responsibility of the business corporation in civil society for many generations to come. Pro-labor legislation alternated periodically with pro-business judicial rulings throughout this period, with labor unions increasingly active in finding new causes to promote. Nowhere was this opportunity more promising than in the mining communities of the West, where Colorado Fuel and Iron (CFI) practiced industrial capitalism in the old-school style. In 1913 alone, 464 men were killed or maimed in mining accidents, and the United Mine Workers of America (UMWA), seizing its opportunity, announced its intention of organizing the state's largest employer, CFI. The strike that followed was a bitter dispute, driving thousands of mine families to seek shelter in tent cities.

After numerous violent exchanges between miners and the company's hired security forces and local sheriffs, the Colorado National Guard was called in. On April 20, 1914, 20 people, 11 of them children, were shot to death by guardsmen. Although the strike was ultimately unsuccessful, the legacy of the "Ludlow Massacre" was to ensure that many of the reforms sought by the UMWA were subsequently introduced. The Rockefellers, long sworn opponents of labor unions, hired Mackenzie King, former Canadian Labor Minister, to create a new approach to management/labor relations. Chernow quotes King's diary entry after a meeting with John D. Rockefeller, Jr.: "Today, there was a social spirit abroad, and it was absolutely necessary to take the public into one's confidence, to give publicity to many things and especially to stand out for certain principles very broadly."

Out of the experience of the Ludlow Massacre came John D. Rockefeller, Jr.'s decision to hire Ivy Lee, who was at that time Executive Assistant to the President of the Pennsylvania Railroad. While Lee was harshly criticized for distributing false information about the wages the miners were being offered and about the past employment of Mother Jones, the union organizer, we can see in his advice to Rockefeller Jr. the outlines of a corporate communications strategy that would be recognizable today. On his counsel, Rockefeller eschewed past practices of buying positive coverage in newspapers and astutely used the theater of the Walsh Commission's investigation into the massacre to begin to repair the enormous damage it had caused to his family's reputation. When it was suggested that Rockefeller enter the hearing chamber through a rear door, Lee insisted that he enter through the same door as the rest of the public. Under Lee's guidance, Rockefeller entered down the center aisle, shook hands with Mother Jones and other union officials, and in response to harsh questioning from Walsh delivered a measured apology that managed to charm Mother Jones even as she strongly rejected his labor relations plan for a company union. On a subsequent trip to the mine in Colorado, Rockefeller agreed to Lee's recommendation that he and Mackenzie King wear cheap overalls on the mine visit over their suits. At a meeting after that visit with employees, Rockefeller spontaneously suggested an impromptu dance and danced with each of the 20 women in attendance.

It was to be another 18 years before the UMWA won recognition at CFI, and two years later Rockefeller and King's form of company union was outlawed by the Wagner Act. Nonetheless, the Rockefeller response to the massacre was to usher in a fundamentally different idea of labor-management relations that definitively established the idea that companies had a responsibility to work together with employees for their mutual benefit.

Ivy Ledbetter Lee's Declaration of Principles

This is not a secret press bureau. All our work is done in the open. We aim to supply news. This is not an advertising agency; if you think any of our matter ought properly to go to your business office, do not use it. Our matter is accurate. Further details on any subject treated will be supplied promptly, and any editor will be assisted most carefully in verifying directly any statement of fact. Upon inquiry, full information will be given to any editor concerning those on whose behalf an article has been sent out. In brief, our plan is frankly, and openly, on behalf of business concerns and public institutions, to supply the press and public of the United States prompt and accurate information concerning subjects which it is of value and interest to the public to know about. Corporations and public institutions give much information in which the news point is lost to view. Nevertheless, it is quite as important to the public to have this news as it is to the establishments themselves to give it currency. I send out only matter every detail of which I am willing to assist any editor in verifying for himself. I am always at your service for the purpose of enabling you to obtain more complete information concerning any of the subjects brought forward in my copy.[6]

During World War I, President Woodrow Wilson established the Committee on Public Information (CPI) to manage news and create support for the war at home and abroad. Mississippi newspaper editor George Creel organized the "Four-Minute Men," an army of volunteers who gave brief speeches wherever they could get an audience—in movie theaters, churches, synagogues, and in union, lodge, and grange halls. Creel later claimed that his 75,000 amateur orators had delivered over 7.5 million speeches to more than 314 million people. CPI publications from the "Four-Minute Men" crusade offered tips on developing and delivering a brief, effective speech.

Committee on Public Information

General Suggestions to Speakers

The speech must not be longer than four minutes, which means there is no time for a single wasted word.

Speakers should go over their speech time and time again until the ideas are firmly fixed in their mind and can not be forgotten. This does not mean that the speech needs to be written out and committed [memorized], although most speakers, especially when limited in time, do best to commit.

Divide your speech carefully into certain divisions, say 15 seconds for final appeal; 45 seconds to describe the bond; 15 seconds for opening words, etc., etc. Any plan is better than none, and it can be amended every day in the light of experience.

There never was a speech yet that couldn't be improved. Never be satisfied with success. Aim to be more successful, and still more successful. So keep your eyes open. Read all the papers every day, to find a new slogan, or a new phraseology, or a new idea to replace something you have in your speech. For instance, the editorial page of the *Chicago Herald* of May 19 is crammed full of good ideas and phrases. Most of the article is a little above the average audience, but if the ideas are good, you should plan carefully to bring them into the experience of your auditors. There is one sentence which says, "No country was ever saved by the other fellow; it must be done by you, by a hundred million yous, or it will not be done at all." Or again, Secretary [William] McAdoo says, "Every dollar invested in the Liberty Loan is a real blow for liberty, a blow against the militaristic system which would strangle the freedom of the world," and so on. Both the *Tribune* and the *Examiner,* besides the *Herald,* contain President [Woodrow] Wilson's address to the nation in connection with the draft registration. The latter part is very suggestive and can be used effectively. Try slogans like "Earn the right to say, I helped to win the war," and "This is a Loyalty Bond as well as a Liberty Bond," or "A cause that is worth living for is worth dying for, and a cause that is worth dying for is worth fighting for." Conceive of your speech as a mosaic made up of five or six hundred words, each one of which has its function.

Get your friends to criticize you pitilessly. We all want to do our best and naturally like to be praised, but there is nothing so dangerous as "josh" and "jolly." Let your friends know that you want ruthless criticism. If their criticism isn't sound, you can reject it. If it is sound, wouldn't you be foolish to reject it?

Be sure to prepare very carefully your closing appeal, whatever it may be, so that you may not leave your speech hanging in the air.

Don't yield to the inspiration of the moment, or to applause to depart from your speech outline. This does not mean that you may not add a word or two, but remember that one can speak only 130, or 140, or 150 words a minute, and if your speech has been carefully prepared to fill four minutes, you can not add anything to your speech without taking away something of serious importance.

Cut out "Doing your bit." "Business as usual." "Your country needs you." They are flat and no longer have any force or meaning.

Time yourself in advance on every paragraph and remember you are likely to speak somewhat more slowly in public than when you practice in your own room.

There are several good ideas and statements in the printed speech recently sent you. Look it up at once.

If you come across a new slogan, or a new argument, or a new story, or a new illustration, don't fail to send it to the Committee. We need your help to make the Four-Minute Men the mightiest force for arousing patriotism in the United States. Committee on Public Information.[7]

The economic expansion of the 1920s was a fruitful time for practitioners of publicity as automobiles, home appliances, and new technologies found ready markets. The works of Ivy Lee, journalist Walter Lippmann, and public relations practitioner Edward Bernays illustrated the increasing interest in the "engineering of consent" not just by appeals to the rational mind through education but through appeals to human psychology and emotion. As the concept of the subconscious and other Freudian precepts became mainstream ideas, the language in which corporate communication was described also underwent a sea change. AT&T's Vail set out to form rational and educational connections with the public. In contrast, William P. Banning, the firm's Director of Public Relations in the early 1920s, sought to make emotional connections just as important. Stuart Ewen quotes from a presentation Banning made to the Bell System Publicity Conference in 1923: "The job of the Publicity Directors of the Bell System is to make the people understand and love the company…to hold real affection for it…to make it an honored personal member of their business force, an admired intimate member of the family."[8]

This sentiment was to be the leitmotif for corporate communications throughout the prewar period. During the 1920s, the benefits of business formation were strongly felt, and it was in this spirit that Calvin Coolidge was able to say that "the business of America is business." The economic expansion of the period was characterized by the creation of the first mass consumer markets in automobiles, household goods, clothing, and entertainment. This was made possible by new techniques in advertising and promotion that utilized the full potential of neon signs, radio, and consumer credit. Whereas buying on installment had previously been regarded as imprudent and déclassé, the creation of demand during the Roaring Twenties swept away these concerns.

Even as banks failed and corporations laid off employees during the Great Depression, companies still sought to project an image that suggested they were operating in the public interest. One of the mysteries of the 1930s, when socialism arguably found its largest voice, is how loyal the American public at large stayed to the ideals of free enterprise. They were aided in this loyalty by the words and actions of the corporations that strove to maintain an image of public service in the midst of bitter economic circumstances.

In the 1940s, corporations worked closely with government to transform the peacetime economy into a fighting machine. For a variety of reasons, the image of the war profiteer never dominated the public imagination as it had during World War I, although Harry Truman, later President Truman, enhanced his image enormously through his work in investigating war contracts during the 1940s.[9]

In the ruins of war-torn Europe and Asia, economic recovery was the sole focus of millions of families devastated by the global conflict. In this environment, the model of the American corporation was the envy of the world. It was as the "ugly American" that U.S.-based multinationals ultimately became known, but consumer admiration was enormous for the industries that produced such advanced technologies as television, refrigerators, and "homes fit for heroes." More significantly, the transition from World War II to the Cold War was so seamless that when President Truman ruled against striking steelworkers, he had the support of much of the nation.[10]

In the 1950s, even as economic prosperity raised the standard of living of the masses, there were warning signs that American multinationals were perceived to be operating high-handedly, even outside the law. In one of his final addresses to the American people, President Dwight D. Eisenhower warned of the threat posed by the "military-industrial complex" to the rights and liberties of the public. Although the 1950s were, in many ways, the heyday of labor unions that negotiated generous wage and benefit increases, there was also a distinct strain of paranoia among commentators about the structure of society and the influence of "big business" that became increasingly widespread. Sociologist C. Wright Mills captured some of this paranoia in his 1956 work *The Power Elite*, which sought to demonstrate that the lives of the many were entirely controlled by a small elite of military, business, and political leaders.

Another author of the period, Vance Packard, built on this concept in his 1957 bestseller *Hidden Persuaders*, which purported to show how mass marketers manipulated the public through psychological research into buying ever-increasing amounts of consumer goods. While his theory that advertisers were flashing subliminal messages in film and television advertising was never proven, the overarching belief that corporations were using insights into human psychology to sell products attained almost universal acceptance.

By the 1960s, fear of big business came to embrace not only the military-industrial complex and marketing manipulation, but the impact of mass production on human health and the environment as well. With the publication in 1962 of Rachel Carson's *Silent Spring* and its widespread support in mainstream circles, the modern industrial corporation found itself under severe attack from a wide range of sources. This newfound skepticism had diverse roots, even as scientific advances continued to produce life-extending and -enhancing technologies. Thalidomide, which was never approved in the United States, was a medication used to counter morning sickness. Between 1956 and 1962, its use created 10,000 children with severe birth defects in Europe and Asia. Agent Orange, used by the U.S. military as a defoliant in Vietnam between 1961 and 1971, was shown early on to create significant health risks for individuals exposed to it.

The emergence of pressure groups, now called non-governmental organizations (NGOs), can be dated to this period in which the contributions of science—and, by extension, the industry exploiting that science—began to be questioned again more broadly than at any time since the Luddites of the 19th century. Friends of the Earth was founded in 1969 and became an international network in 1971, the same year that Greenpeace came into being in Vancouver. Corporations reacted initially by claiming that the research detailing environmental harms was faulty. They also engaged in other unappetizing behavior, such as ad hominem attacks on Rachel Carson, claiming that, as a woman, she was unqualified to opine on matters of science. However, by the end of the 1960s, the evidence for environmental harm became overwhelming. The U.S. Environmental Protection Agency opened its doors in December 1970 and in the ensuing decade oversaw the implementation of clean air and water acts, the banning of DDT and aerosol fluorocarbons, and the removal of lead from gasoline.

The 1970s can be viewed in hindsight as the era in which public opinion about the motives and ethics of big corporations migrated out of the publications of the political left into the mainstream media and into mainstream opinion. Arguably, the disenchantment of the general public with the behavior of corporations cannot be disentangled from its view of the political establishment. Even for conservative Americans who were outraged by opposition to the war in Vietnam, the events of Watergate that unfolded between 1972 and 1974 were to prove a bruising awakening.

For the Watergate generation of congressmen, as well as for the newspaper reporters who burned to emulate Woodward and Bernstein, there was apparently plenty of corporate misbehavior to attack. Nor did events necessarily give the lie to their conviction that corporations were in business to make money irrespective of the human consequences. Reliably, a major scandal erupted every three to four years throughout the 1970s and 1980s, providing fodder for anticorporate apologists and pressure groups. A selected chronology of the era shows us:

1971–1974	Dalkon Shield IUD renders thousands of women infertile
1976	Explosion at Seveso, Italy. 80,000 animals exposed to dioxin slaughtered
1978	Residents at Love Canal link health problems to buried toxic waste
1979	Three Mile Island nuclear plant comes close to meltdown
1984	Explosion at Union Carbide in Bhopal kills 8,000 people
1986	Nuclear explosion at Chernobyl requires evacuation of 336,000 residents
1989	Exxon Valdez accident dumps 11 million gallons of crude oil into Prince William Sound

In addition to these distinct newsmaking events, it should not be forgotten that the period between 1969 and 1985 was more challenging economically than any period since the end of World War II. In this environment, popular antipathy to business found a ready audience in a population wrestling with what became a deadly combination of inflation and stagnation, or "stagflation." In fact, by 1979, a study by pollsters Rothman and Lichter showed that more than 20 percent of employees of *The New York Times* believed that major corporations should be nationalized. Chief villains in this morality play were not surprisingly the multinational oil companies for whom the global oil crisis was a period of intense vilification. To many industry opponents, the Exxon corporate logo represented "the company with the double cross." By the end of the 1960s, widespread antipathy to the industry had become so entrenched that many corporate communication executives wondered whether the public corporation would ever regain the standing it had enjoyed in the 1950s.

One such executive who decided to do something about it was Herb Schmertz, then the most senior communications executive at Mobil Oil. Schmertz's perception, as he described later in his book *Goodbye to the Low Profile* (1986), was that American corporations in general and oil companies in particular had lost the confidence of their own convictions and, by failing to engage effectively with opponents, were contributing to their own worsening position. His solution was to commence a corporate communication campaign based on a recurring series of editorial advertisements, dubbed "advertorials" which ran for many years on the op-ed page of *The New York Times*. Each placement contained nothing more than copy from Mobil addressing an individual issue with which the company and its stakeholders were concerned. These included topics such as environmental protection, the price of oil, and the question of remaining oil reserves, among others.

The insight underpinning this campaign was that if the company's opinions about issues affecting it were not being aired in other media, then the company owed it to its stakeholders to ensure that they were aired somewhere. Furthermore, by conducting an extended campaign in this manner, Schmertz demonstrated that the modern corporation benefited from having a visible public campaign supporting its viewpoints that was not merely intermittent or confined to one specific issue but that became an ongoing component of a continuous effort. Corporate brand communication efforts of today, such as the "Ecomagination" platform of General Electric or the "Human (Hu) Element" advertising of Dow Chemical, owe their existence to the inspiration of the Schmertz advertorials.[11]

Even after corporations stepped back from the kind of character assassination and ad hominem attacks to which they had subjected Rachel Carson and other early environmentalists, they struggled to find an appropriate tone of voice in which to deal with their adversaries. The sharp economic downturn of 1982, for example, notably provided filmmaker Michael Moore with abundant material to skewer corporate communication in his film *Roger & Me*. This popular film, which examined various impacts of the economic crisis seen through the lens of Moore's leftist consciousness, was threaded on his failed attempts to get an on-camera interview with General Motors chairman Roger Smith. Through skillful editing, Moore managed to depict the corporation's communication team as a modern-day version of the Keystone Kops in its bumbling attempts to shield company executives from hostile questioning.

If the Schmertz "advertorial" campaign represented a deliberate innovation to give corporations back their voice, another event was inadvertently to set a new standard for corporate behavior. On September 30, 1982, a reporter for the *Chicago Sun Times* called Johnson & Johnson asking for information about the manufacture of Tylenol. So began the events that became known as the "Tylenol Crisis," in which seven people died after ingesting Tylenol capsules that had been laced with cyanide. Even though the company was able to determine relatively quickly that the tampering had not taken place inside one of its plants, on October 5, CEO James Burke announced a nationwide recall of all Tylenol products at an ultimate cost of more than $100 million. The perpetrator was never caught. This incident is remembered best for the extraordinary decision that Johnson & Johnson took, and also for the forthright and uninterrupted communication that flowed to consumers from the company throughout the crisis. By November, the company had devised new tamper-proof packaging that has become the industry standard and regained 100 percent of its pre-event market share.

James Burke and the team at Johnson & Johnson have rightly received great praise for their recall decision and the communication effort that surrounded it. Less frequently commented on is the fact that their decision to recall the entire product line fundamentally shifted public perception about the responsibility of a corporation to its customers. In the Tylenol crisis it was quickly established that neither the company nor any of its employees were responsible for the poisoning of the capsules. The FBI and the FDA initially counseled against the recall on the grounds that it would give inspiration to terrorists. Nonetheless, by recalling Tylenol products, Johnson & Johnson adopted the view that protecting its customers was its most important goal, irrespective of cost and irrespective of liability. By doing so, this New Brunswick, New Jersey-based corporation opened a new chapter in the history of corporate accountability that still reigns supreme today.

Notes

1. Timothy Penning, "First Impressions: US Media Portrayals of Public Relations in the 1920s," *Journal of Communication Management* 12, no. 4 (2008): 358.
2. John Steele Gordon, "The Public Be Damned," in *The Business of America* (New York: Walker & Co., 2001), 96–99.
3. CCI Practices and Trends Studies, **www.corporatecomm.org/studies**.
4. Jack Beatty, *The Age of Betrayal* (New York: Random House, 2007).
5. Chernow, Ron. *Titan: The Life of John D. Rockefeller, Sr.* (New York: Random House, 1998).
6. See Karla Gower, "US Corporate Public Relations in the Progressive Era," *Journal of Communication Management* 12, no. 4 (2008); 305–18; Karen Russell and Carl Bishop, "Understanding Ivy Lee's Declaration of Principles," presented at Association for Education in Journalism and Mass Communication, Chicago, August 2008, www.allacademic.com/meta/p272004_index.html; Ray Hiebert, *Courtier to the Crowd: The Story of Ivy Lee and the Development of Public Relations* (Ames: Iowa State University Press, 1966). While Lee's Declaration of Principles is an important founding document for the corporate communication discipline, subsequent research has shown that his news reports on the Ludlow Massacre were not consistently accurate and his representation of IG Farben and the Japanese railroads in the 1930s remains controversial.
7. *Four Minute Men Bulletin* 1, May 22, 1917.
8. Stuart Ewen, *PR! A Social History of Spin* (New York: Basic Books, 1998).
9. David McCullough, *Truman* (New York: Simon & Schuster, 1992), 253ff.
10. Ibid., 280–81.
11. Herb Schmertz and William Novak, *Goodbye to the Low Profile: the Art of Creative Confrontation* (Boston: Little, Brown, 1986).

CHAPTER EIGHT

Philosophy

The Engineering of Consent and Process: Strategic and Tactical Models

There is an almost infinite number of job titles and job descriptions covering the role of corporate communication in large organizations. However, as a practical matter, a number of strategic and tactical themes can be identified into which these various descriptions can be grouped.

Strategic Models

Strategically, the function of corporate communication can be described as serving these corporate goals:

- Advocacy or the engineering of public opinion
- Stewardship
- Counsel

Depending on the structure and history of an individual organization, as well as the communication needs of a particular industry, these three goals are achieved through an individual or single unit or separate structures and functions within the organization. In addition, the importance ascribed to these different corporate goals will vary depending on numerous factors such as the business cycle, the policy climate in an individual country or industry, and the maturity of the communication function in an individual organization. Finally, initiatives undertaken in support of any of these three goals are rarely hermetically sealed. Advocacy activities, for example, will frequently benefit the corporate brand stewardship efforts. Effective counsel to senior management becomes an integral ingredient in both advocacy and stewardship. We treat them separately here in order to facilitate a clear discussion of how corporate communication supports these goals.

The Advocacy Model

Strategically, advocacy—or the engineering of public opinion—takes place on a variety of levels. We can break these down into some important constituent parts:

- Direct advocacy with decision-makers in support of or in opposition to existing or proposed regulation or legislation at a global, regional, national, or provincial level
- Influencing public opinion in support of a company's or industry's viewpoint on a specific issue
- Shifting public opinion over an extended period of time with respect to broader philosophical issues, such as the extent of government regulation, the structure of the tax system, and the balance between public and private participation in the marketplace

The advocacy model is, perhaps, the easiest to understand and to explain. Since its beginnings in the 19th century, the modern corporation has existed in national, state, and provincial public policy environments. Its ability to thrive in a commercially viable manner has depended on its advocacy of policy positions that benefit its growth and opposition to policies that would inhibit that growth or even cause the corporation to shrink or become less profitable. Throughout the past 150 years, the range of regulatory and legislative issues for which this advocacy has been required has been enormous. Among the most contentious early policy disputes after the Civil War was the subject of tariffs imposed on imported goods. "Twenty men can enter a room as friends," Will Rogers joked, "and someone can bring up the Tariff and you will find nineteen bodies on the floor with only one living that escaped."[1] The same kinds of advocacy conflicts have been conducted on the issues of granting land to railroads, defending the trusts from President Theodore Roosevelt's reforms, for and against Prohibition, the minimum wage, the securities laws that followed the stock market crash, and almost every piece of legislation affecting corporations up through climate change and the carbon footprint. It can fairly be said that these advocacy battles have formed an underlying and arguably fruitful tension between the rights and responsibilities of the corporation and those of other participants in the modern capitalist system. Through different societies and different eras, these disputes have been conducted in both ethical and in corrupt ways, causing the political establishment in those different societies to impose greater or lesser restrictions on the ability of companies to advocate for their interests. By and large, however, the fundamental right of companies to do so has been an enduring feature of democratic and capitalist societies over the past century and a half.

Examples of these three core strategic communications efforts come readily to mind. In 1994, a consortium of U.S. companies with global business interests and with the support of the Business Roundtable and the U.S. Chamber of Commerce sought Most Favored Nation status for the People's Republic of China, a status that had been suspended in 1951 by President Harry Truman. Passage of this legislation eased the ability of U.S. companies to do business with and in China, but it was vigorously opposed by companies with a reason to fear imports from China, by labor unions concerned about job losses, and religious and human rights groups opposed to the Chinese government's policies. The campaign was conducted primarily by direct lobbying of legislators but was also accompanied by communications directed at influential persons in the media and the academe.

We also have a well-known example of the second type of campaign from the 1990s. In this case, the struggle concerned health care reform and is distinguished from the first type of strategy to the extent that it was mass public opinion, as opposed to influential elites, who were the target of the campaign. In 1993, President Bill Clinton proposed legislation closely associated with the chair of the taskforce he appointed, Hillary Clinton, to put in place universal health care insurance. A spectrum of participants in the health care industry, fearing that this legislation would damage their businesses, mounted a large-scale campaign to persuade the public to voice its opposition to what was quickly dubbed "Hillarycare." The legislation was ultimately defeated. Credit for the defeat was given to the public opinion campaign waged by the coalition against it, most notably a series of television commercials in which an average American couple, "Harry and Louise," questioned whether health care should be run by government bureaucrats.[2]

The third type of advocacy has a more contentious legacy but is still regarded as a legitimate exercise of the advocacy function. This type of advocacy generally takes place at the philosophical level and, while it generally aims ultimately to achieve regulatory or legislative outcomes, it takes place in the marketplace of ideas rather than that of lawmaking. In this category we can count the long-running efforts by conservative think tanks, supported in some cases by corporate foundations, to change the overall tenor of public policy in certain areas. Tort reform, the effort to make it more difficult to sue corporations or to recover large sums in damages, probably represents the most prominent and longest-running example of this type of advocacy. Over the past 20 years, advocates of changes in laws governing litigation have invested significant resources at federal and state levels to create a body of thought sympathetic to their aims. The principal vehicles for this effort have been the creation and funding of think tanks to produce academic research papers and conferences dedicated to the sponsors' issues, the establishment of university chairs with an ideological focus, and the support of longitudinal research to show long-term impacts. While there are examples of such advocacy on the part of liberal and left-leaning organizations, the majority of such advocacy has tended to come from ideologues supportive of corporate "rights."[3] The reason why some of this advocacy has been controversial is because critics argue that it is a poorly disclosed, even secretive effort to influence public opinion by funding research designed to look and sound neutral and objective in the public eye. The sensitivity of various audiences to this critique is a reflection of the much-debated but unresolved question of how funding influences scientific and academic objectivity in democratic societies.

The Stewardship Model

The variants of the advocacy model discussed above concern themselves with supporting the growth and nourishment of an organization around specific issues. In the stewardship model, we are concerned with the role of the corporate communication function in protecting and promoting the good reputation of the organization with its stakeholders—employees, shareholders, customers, business partners, communities, and regulators, to name the most important. This function has often been described as safeguarding the reputational capital that an organization builds up over time, or as a "trust" account which needs to be protected. These somewhat hackneyed metaphors are not without merit. The good opinion held by stakeholders does in fact correspond in many ways to the accumulated good acts and strong relationships that a well-managed organization enjoys. Similarly, when a company encounters a reputational challenge, it is said to be able to "draw on" its stored up good will. Stakeholders will give it the benefit of the doubt and, in the absence of concrete evidence to the contrary, will ascribe good faith to the company's actions.

The problem with this particular kind of imagery is that it obscures the fact that brand stewardship, to extend the metaphor momentarily, is not about checking on the monthly bank statement to make sure that the reputation account is in good standing and then filing it away. Effective corporate brand stewardship requires an active, ongoing, and continuously refreshed program to ensure that the organization presents itself to the world and its stakeholders in a way that is aligned with the changing global environment. Two brief examples will illustrate the dynamic approach required to ensure that a company does not fall out of step with changing expectations for good corporate behavior.

The first example derives from the world of crisis communication. Through the 1970s, communication with stakeholders in a time of company crisis was deemed to be one of the principal jobs for the company spokesperson. It was rare to find examples of CEOs taking a public role during the acute phase of a crisis. Indeed, one could fairly say that senior executives of public corporations didn't feel an obligation to present their view to the public in the way that has become the norm today. Starting in the 1980s, however, and motivated to some degree by the poor performance of a number of highly visible companies in communicating during crises, visible CEO leadership became the hallmark of good communication practice. Furthermore, strong performance by the CEO and other senior leaders during crises came to be regarded in and of itself as a marker for well-run and respected companies.

Our second example comes from the field of environmental stewardship. In the 1960s, and especially in the 1970s, companies in the vanguard on environmental issues created programs to demonstrate that they were committed to remediating past failures to protect the environment and to clean up toxic sites and polluted waters. The corporate standard bearers for good behavior toward the environment developed a narrative of harm reduction that became the gold standard of environmental stewardship. If we fast-forward 25 years, however, we see that public expectations about what constitutes good stewardship of the environment have changed substantially. Under the banner of corporate social responsibility, itself a term that connotes a changed relationship of corporation to society, environmental stewardship has come to mean much more than harm reduction or remediation. Leadership in the environmental sphere in this millennium has come to mean investing proactively in strengthening the geophysical environment of the planet and transforming one's operations to create the smallest feasible environmental "footprint."

What these two examples show is that unless corporate brand stewards are actively engaged with stakeholders and constantly adapt the brand and its communications to changing philosophical and ethical expectations of those stakeholders, its reputational capital will be eroded. In order to combat this reputational entropy, communicators need to be hyper-attuned in these three areas:

- Issues monitoring and management
- Crisis communication response
- Corporate Citizenship and Social Responsibility/Sustainability

Issues Monitoring and Management

The phrase "issues monitoring" often invokes an image of teams of junior staffers punctiliously gathering all references to a company and its industry that flow daily through the many print and virtual media channels and neatly packaging them for management perusal. This is indeed an important activity to which we will return later, but there is a fundamental strategic orientation that needs to precede and enfold the tactical monitoring activity. An issues monitoring and management strategy needs to be based on a process we call "issues mapping." The essence of this approach is to chart all of the principal activities in which a company is engaged as a business and then mapping these activities against changes taking place in the external global environment. In order to create a taxonomy for these changes and help prioritize them, we have found it useful to divide these external changes into four forces:

- Demographic
- Technological
- Socio-cultural
- Economic

We can perhaps best illustrate this approach by using a medical diagnostic testing company as an example. The core businesses of such a company would normally involve extracting a tissue/blood sample from an individual, transporting samples to a testing facility, recording and communicating the results to a physician, storing and/or destroying the sample, and, finally, processing payment through a health insurance provider. How might the four forces operate on such a business? The aging of the population of the developed economies (demographic force) could produce a customer base in which a larger percentage of people the company is testing are on multiple medications or could be immuno-compromised.

How does this impact a company's pre-test questionnaire and the safety procedures in its test-taking? With the increase in digital record-keeping (technology force) and the explosion in cybercrime, will the company continue to upgrade its computer security to cope with new threats? As the population the company serves becomes more diverse (socio-cultural force), will the gender mix in its recruitment meet the needs of female customers who will only be tested by a woman? The company's van drivers are low-wage employees with high school diplomas. In a thriving economy (economic force), these employees can get better jobs, leaving the company with a less educated pool. Do the company's training programs correct for this shift?

What emerges from this audit should be a clear picture of how a company's operations "map" against changing conditions and expectations. It provides a clear picture of those areas that could represent reputational potential but most importantly helps unearth areas that could pose reputational threats. It is noteworthy that this aspect of brand stewardship does not primarily involve communication but applies a reputational filter to operational choices that companies need to make as they conduct their business.

Alongside this systemic analysis, corporations need to develop an ongoing system for tracking issues that could affect their reputation. These are issues that may be specific to a particular company or industry sector, but they can also be issues revolving around corporate governance, diversity, and sustainability that any well-managed company should be following on a daily or weekly basis. For diversified and global companies, it is increasingly important to stay up to speed on issues of geopolitical risk. While it may appear that the changing threat pattern for energy companies in West Africa, for example, might seem to be a concern for operations, not the communication department, the current "standard of care" is increasingly to expect corporate communicators to have a viewpoint or position on any issue that has the potential to impact the company. For some kinds of issues, such as particularly thorny matters of international trade or taxation that don't directly touch a company's business, it is still important to be able to demonstrate awareness of the potential impact in order to convey the kind of vigilance that shows strong leadership. For issues on which a company could reasonably be expected to have a position, it is crucial to keep a Q&A log that is continuously updated as circumstances change.

Crisis Communication Response

As described earlier, theories of crisis communication have evolved significantly over the past 30 years. In the 1960s, it was still reasonable to suggest that a developing crisis be kept confidential until the "problem" had been solved by the corporation. By the 1990s, the public's right to know about a crisis immediately became a firmly established principle, and companies that were able to communicate early and often about an emerging crisis reaped significant reputational benefit. Swift preemptive action also became a hallmark of effective crisis response as the Johnson & Johnson Tylenol poisoning case in 1982 demonstrated.[4] As the literature on crisis communications has grown over the past decade, it is also increasingly clear that in many cases bungling the communication connected with a crisis is as significant from a reputation capital point of view as the fallout from the crisis itself.

With the arrival of widespread broadband access, the explosion in cell phone use, and the emergence of social networks, it has become clear that crisis communication now needs to be conducted at an exponentially higher level of transparency. This reflects not only the fact that the tools of Web 2.0 themselves dramatically speed up the so-called pro-dromal phase of a crisis (the initial phase in which early symptoms begin to appear) or that crises are increasingly "net-generated" events in themselves, but that the multiplicity of communication channels on the social web creates an expectation of outreach to all stakeholders by the affected organization. Since these tactical delivery mechanisms cannot be constructed from a standing start in the midst of a crisis, corporations are increasingly pre-staging crisis response mechanisms, so-called "black" or "dark sites," that can be activated with the click of a mouse in the event of an actual crisis.

Strategically, crisis communication has also become more complex for a number of other reasons. The principal development is that Web 2.0 has broken down the barriers between what one might call ordinary information consumers, experts and activists who can now access each other's viewpoints and content with a simple web search. Even if the average information consumer would ordinarily lose interest in a given situation relatively quickly and allow an issue to die down, in today's RSS (Really Simple Syndication) streaming world, that consumer has to actively disconnect from the feed informing him about the dialogue a corporation might still be having with the experts or the activists. This turn of events puts enormous pressure on companies trying to decide how, where, and for how long to continue communicating about a crisis. The leading practice in this area is in the process of being established, but we are able to discern the outline of some guiding principles as articulated in Figure 8.1.

As companies have discovered at some cost, the deployment of these web-based and mobile communications tools requires a high degree of authenticity and a consistency of tone that is not always easy to achieve. The term authenticity has, to some degree, become a buzzword of corporate communication. It is the term underlying an excellent White Paper issued by the Arthur W. Page Society in 2008 called *The Authentic Enterprise*. The authors of the paper seek to describe the values orientation of the contemporary corporation. We use the word in a more circumscribed sense to mean that the behavior and utterances of an organization in a crisis need to be identifiably consistent with past behavior and utterances.

Communications Channel	Type of Engagement
Company Website	Useful when the issue or crisis is ongoing, when information needs to be updated frequently, and when the facts or the company's position are distorted by critics.
Corporate Blog	Once up, the corporate blog has to stay up with no less than weekly postings. Useful for ongoing issues and to explain complex issues in simple terms.
Social Media Platforms	Requires a long-term commitment and the willingness to cede some control. Avoid heavy "corporate" tone. Not suitable for crises involving physical harm.
Activist Blogs	Suitable for high-temperature issues for which the company is also engaging in other media. Expect to be "disrespected" and engage them offline, too.
Micro-Blogs (Twitter)	Useful for rapidly evolving events to share information quickly and broadly. Assume other participants are also "tweeting" from the scene.

Figure 8.1 Guiding Principles for Crisis Communication on the Web

It is this sum of past experiences that stakeholders have with an organization that creates what we call the "baseline of expectations." This baseline can differ radically, depending on a variety of factors such as the pre-existing visibility of the corporate brand and how proactively that brand has painted itself as being engaged with and responsive to all of its stakeholders. For example, BP has suffered reputational damage to a disproportionate degree from recent events precisely because its "Beyond Petroleum" corporate campaign appeared to hold the company to a higher ethical standard than other energy companies. By contrast, the management of ExxonMobil has been more parsimonious in its communications and in raising consumer expectations about the future of alternative energy.

At a strategic level, then, the modern corporation seeking to protect its reputation in a crisis needs to embrace the concept that reputation is driven by operations as much as statements and, therefore, needs to understand where the reputational threats in those operations can come from. Furthermore, the speed and complexity of channels available to stakeholders requires detailed advance planning so that the company's response can appear timely in today's hyperfrenetic information sphere. Finally, the mode of engagement and the tone of the engagement that an organization chooses in a crisis must be consistent with its professed values and the baseline of expectations it has established with stakeholders.

Corporate Citizenship and Social Responsibility/Sustainability

Corporate Citizenship, Corporate Social Responsibility, Sustainability: three different names for what one could arguably call the central tension in contemporary society's view of what corporations need to do to deserve a license to operate. At either end of the ideological spectrum, there is very little to argue about. Milton Friedman, the Nobel Prize-winning economist, wrote a much-quoted essay in *The New York Times Magazine* in 1970 in which he articulated the position that "the social responsibility of business is to increase its profits."[5] In Friedman's view, any attempt to make business decisions on the basis of any other criterion was simply taking money out of the pockets of shareholders, employees, or customers. (See the discussion in Chapter 4 above, "Strategic Ethical Relationships: Trust and Integrity.") To the socialists and crypto-socialists of the 1960s, the private corporation existed purely at the sufferance of the rest of civil society and could be directed through taxation and regulation to provide whatever social goods were desired, largely irrespective of commercial logic.

In between these two extremes, however, there has been a rich debate about the extent to which businesses should invest in activities not principally designed to create profit but to provide environmental benefits or improvements in the workplace and in the sphere of human rights.

Apologists for CSR are quick to supply evidence that investing in the kinds of activities typically grouped under this heading not only benefit society but enhance a company's profitability. Indeed, there are some clear examples, such as in the reduction in packaging of consumer goods, in which both environmental benefits and costs savings can be found.

However, the confusion to which many companies find themselves subjected is caused by the fact that there is no consensus as to whether CSR activities can only be justified if they do contribute to profitability, or whether the mandate to be "socially responsible" exists even if such activities are a drag on profits. This question is further complicated by the concerns of some critics who point to certain CSR initiatives, such as providing free cell phones to farmers in developing countries, as a thinly disguised marketing program to build infrastructure prior to capturing profitable network usage fees. We will leave for others the question of whether a program that benefits farmers *and* benefits the communication provider is somehow morally ineligible for CSR status, but this example does illustrate the anxiety created in this emerging field.

How then should a company go about devising a corporate social responsibility strategy, and how should it be communicated? We believe it is very important for senior management to begin by asking themselves the tough question of whether they want all of their CSR initiatives to be revenue-generating or whether there is a reputational benefit to these initiatives that will create an important non-financial return. It is the very clarity around this issue that has helped General Electric, for example, to create a powerful campaign called "Ecomagination" around its environmental CSR goals. In typically tough-minded fashion, GE leaders have insisted that all the environmental initiatives undertaken by the company generate revenues and ultimately profits. This clarity has enabled the firm to communicate very effectively with stakeholders from investors to employees to consumers.

It has also helped create internal benchmarks for success that other companies have struggled with. The apparel and footwear company Timberland has achieved a similar clarity via a very different route. Here is the company's position:

> Timberland's commitment to corporate social responsibility (CSR) is grounded in the values that define our community: humanity, humility, integrity and excellence. For over 30 years, "community" has been synonymous with the ethic of service—the desire to share our strength for the common good. Our approach to building and sustaining strong communities includes civic engagement, environmental stewardship and global human rights.[6]

For Timberland, the company itself is defined by its commitment to the "ethic of service." Employees, shareholders, and customers understand the company's position. It has set for itself bold CSR goals, and its decision to increase the frequency of its CSR reporting to quarterly updates is a powerful symbol of that commitment. What management is saying through this decision is that it is just as important to report quarterly about sustainability as it is about revenues.

Once an organization has framed its own basic position on CSR, it becomes possible to make other strategic decisions that will lead to effective CSR communications. Experience suggests that CSR initiatives and stakeholders' response to them are most positive when the initiatives are closely aligned to the company's core businesses in a clearly definable way. Companies, for example, that base significant revenues on low-wage manufacturing in emerging economies rightly place a major emphasis on human rights in the workplace. Technology companies are well served by a focus on education and innovation. Financial institutions have frequently been leaders in micro-finance and entrepreneurship. Depending on the national regulatory scheme under which the company operates, there may also be mandatory reporting requirements in areas such as the environment or human rights, and most large companies will have some kind of activity in these areas. However, we believe that credibility and respect are created by a strong focus in one or two principal areas.

By establishing a strong strategic framework, it becomes possible to embed corporate social responsibility into the everyday business activities of the corporation and to align incentive structures to reflect this. Once this is achieved, companies can communicate organically about their CSR endeavors on an ongoing basis, minimizing the impact of critics who argue that the glossy annual CSR report is little more than corporate PR without commitment or substance. A well-run CSR effort that delivers what it promises is a powerful example of good brand stewardship.

The Counsel Model

The third core responsibility owned by the communication function in managing public issues goes by the rather bland name of counsel. But contained within this responsibility is a role without ownership elsewhere that is critical to the effective functioning of successful organizations. While there is some overlap with the other key responsibilities for advocacy and stewardship, the counseling function goes beyond either of the other two in embracing the needs of the whole organization, and without it neither of the other two roles could be successfully performed.

In order to explain this role adequately, it is necessary to construct a thought experiment in which we treat an organization as if it were an actual human being with a personality, as opposed to an organization with thousands of human participants. We are not embarking here on a discussion of the long-running legal question of whether a corporation is a "person" in a juristical sense, but rather trying to suggest that, like a person, organizations have both a sense of self and a reality in the perceptions of its stakeholders.

The sense of self that exists in individuals has many aspects to it: How happy am I? How successful am I? Is life treating me well right now, or am I in the doldrums? Are things getting better or getting worse? Am I hopeful or pessimistic? At the same time, the outside stakeholder observer also has a sense of the institution: Do I think this company is helpful? Arrogant? Failing? Successful? Selfish? Lacking in confidence? These perceptions are, of course, shaped by many different inputs, ranging from personal experience to media coverage to financial outcomes. Not surprisingly, since the life of organizations, like the lives of individuals, is dynamic and occasionally volatile, these internal and external perceptions are in constant danger of becoming misaligned. It is the role of the corporate counselor to identify these moments of misalignment and propose activities to remedy them or, conversely, to recommend the cessation of activities that may themselves be causing the misalignment.

How do these misalignments come about? One of the principal causes, generally perceived to be a good thing, is rapid growth. The entrepreneurial start-up with five, then 30 employees, suddenly finds itself three years later with 500 employees and an increasingly dominant share of a significant market segment. This kind of a company is frequently dealing with what we think of as the problems of corporate adolescence. Some functions are highly mature, while others are barely out of infancy. The technology company that is growing rapidly, for example, may have a powerful and sophisticated group of engineers, along with an exploding sales force. At the same time, it has a human resources department sized for a company one-tenth the size and corporate governance practices adequate for a venture-backed start-up rather than a substantial business operation. These failures in business operations are frequently the cause of the misalignments in perception that are so damaging to corporate reputation. However, it is the misalignment between internal and external perceptions that can cause the biggest problems.

To illustrate how the gaps between internal and external perceptions increase over time, let's look at a high-technology company that we'll call "Heliosaurus Enterprises."

In the early days, Heliosaurus is run by a charismatic and driven leader and a dedicated and loyal team that lives out of Chinese food cartons while developing its core technology. They see themselves as unheralded outsiders, taking on the Goliath companies in their market space with better products and phenomenal customer service. They believe themselves to be liberating clients from the evil domination of the market leader, fighting tooth and nail for every sale. In the beginning, this is true, and external stakeholders, customers, analysts, and the media also applaud this brave little upstart.

Five years later, the upstart is the market leader and has hundreds of employees with no shared history with the founder and his inner circle. The direct personal relationships that existed between the first customers and the management team are gone, buried underneath CRM reports and revenue-per-customer targets for the expanding sales team.

Suddenly, the boyish macho culture that brought Heliosaurus such great success seems less charming and more arrogant. The "light a fire under them" tone of the CEO/founder's employee communications seems less inspirational and more like bullying. The technology vision that caused shareholders to put their faith in the management team seems less prophetic and much more like hype. Left unchecked, these misalignments lead unfailingly to tense customer relations, declining employee morale, and a skeptical shareholder base. Indeed, if a company grows large enough in its market, even regulators and lawmakers start to dislike what they see.

We've hypothesized a technology start-up to illustrate these principles, but organizations of all sizes in every field are subject to the power of reputation entropy whether they are two or two hundred years old. It is the ability to identify these misalignments and the credibility to recommend changes to management based on this identification that mark the gifted corporate counselor. Such a person brings many skills to the table, but a key attribute is the strength of character and insight to be able to remove the institutional blinkers that too frequently characterize how leaders see the world. The counselor understands not what a leader thinks people hear when she speaks, but what they actually hear. The counselor, through deep research and observation, detects when the internal and external perceptions of an issue have become decoupled.

There are, of course, many leaders who themselves have an instinctive gift for grasping how these misalignments need to be corrected. Lou Gerstner, the legendary CEO of IBM who took over at the nadir of the company's fortunes in 1993, famously quipped when asked to explain his vision for the company: "The last thing this company needs right now is a vision."[7] His comment instantly captured the gap between IBM's big-company rhetoric and stakeholders' increasing cynicism and cleared away the obstacles to recovery represented by this misalignment. For most, however, the drive, commitment, and focus that is required to lead large organizations makes this empathy and insight hard to achieve. This is why the role of corporate counselor is so crucial in large organizations.

Tactical Models

In fulfillment of the strategic goals of providing advocacy, stewardship, and counsel, the communication function needs to avail itself of several tactical models of execution: company publicity, company reputation, employee relations, public information about the company, and investor relations. Communication channels have grown, some more than others, but the way in which companies engage their constituents, industry colleagues, community, and stockholders has changed dramatically.

Company Publicity

Simply put, company publicity is the extent to which an organization's activities are made known to its stakeholders. These activities stem from operational decisions such as the opening and closing of facilities, the introduction of new products, the funding of research innovation, corporate social responsibility, or the appointment of key executives. Although creating a tactical model for company publicity can be straightforward, it becomes much more complex in large-scale global organizations in which the priorities, processes and personalities of corporate divisions don't always layer up neatly into a cleanly articulated corporate identity.

Take Telecommunications Company X, for example, whose profitability is largely based on revenues derived from its high-value software patents. At the same time, the handset division produces significant top-line revenue, but global competition has all but killed the profit margins. From the CEO's perspective, the handset division is a commodity franchise, and the sooner he can find a buyer, the better. The president of the handset division, whose job is to increase its success, is determined that the company's corporate positioning should emphasize the centrality of handset production to its core business activities. It is not hard to imagine how difficult it is to reconcile these two visions in efforts to design company publicity. These conflicts are magnified exponentially when varying geographic priorities, caused by differing marketplace maturities, come into play.

In addition to this tension, efforts to create effective company publicity are also beset by innumerable calendar conflicts as different parts of the organization seek to exploit seasonal and event-driven opportunities to tell a positive story about their activities. Even when such positive opportunities do not conflict, company publicity efforts are often derailed by regulatory disclosures of a negative nature: the filing of a lawsuit, the termination of a senior employee, or the closing of a facility. While some of these conflicts are unavoidable because of mandatory filing timelines, it is all too common to find one negative aspect of a company's business overshadowing important positive efforts in another.

The third systemic problem that a successful tactical model for company publicity needs to address is the fact that, in most diversified companies, one division is inevitably in a more dynamic part of the cycle than another. To borrow a cliché from corporate-speak, one division will be "shooting the lights out" while others are performing unspectacularly or even underperforming. In this common scenario, it is inevitable that corporate communicators will overconcentrate their company's publicity efforts on the division with a stellar performance. At the same time, the CEO's speeches and conversations with analysts will over-emphasize the performance of this division. Many leaders succumb to the temptation to talk about the over-achieving unit as the "real future of the company." The problems that this creates are manifold, both from a company publicity point of view and, arguably, from an operational perspective.

The first of these difficulties lies in the fact that the cycle almost always does turn, and the star of today hits a plateau or even a dip in fortunes. Having singled out this division as the "future of the company," it then becomes necessary to reverse field, leading to dented credibility for the leadership and a muddled picture of the company's true capabilities. An additional casualty of having made a star of the growth division is the resentment of other divisions, who in many cases have been delivering solid—if unspectacular—value to customers but whose products may be less glamorous in the eyes of company leadership or the media. This has been a particular feature of the Internet era. In company after company, the unprofitable but soaring web-based business unit crowds out all other activities of the company's publicity effort. When, as in almost all cases, the business model for this initiative fails to vindicate itself, the ensuing about-face sets back the ongoing publicity campaign significantly. Going cap in hand to the other, less-glamorous divisions, the red-faced communication team frequently encounters a less than embracing divisional management.

Finally, as has been suggested, there may even be an operational reason why playing favorites among the company's divisions from a publicity perspective can have a deleterious effect. The original sin, as we might describe it, is committed when the corporate communication team neglects to solicit ideas and case studies from divisions perceived to be unglamorous. Not only does this create resentment, but it can also cause the "stodgy" businesses to retreat into their shells and stop thinking creatively about the positive stories that they do have to tell. This kicks off a vicious cycle in which, for lack of airtime, promising initiatives in these divisions fail to receive attention, particularly the kind of attention that begets funding. Initiatives starved of funding die. Businesses without innovation initiatives tend to underperform and, as a result of this cycle, remain trapped in the doldrums. They shrink and are often sold off as a result.

The reason for dwelling on this last point is to emphasize the fact that company publicity is far from a vanity project undertaken for the company or its leadership. To the extent that it creates external visibility for the company's activities, it stimulates customers and other stakeholders to want to interact with it. To the extent that it creates internal visibility for worthwhile projects, it unleashes the creativity and energies of those engaged in that initiative, and pushes others in the business unit to develop newer projects in the hope of basking in the same internal and external glow created by the first initiative. The evolution of peacock-feather displays grew by emulation. Likewise, company publicity, equitably distributed, grows through innovation and business success.

One conservative company that has learned this lesson well is General Electric. In 2003, it launched a company publicity initiative called Eco-Imagination, whose message was twofold: (1) GE had the engineering skills and the creativity to bring to market breakthrough products that would lessen the human burden on the planet; (2) GE would make money by engaging in this endeavor. The effect of this campaign was electrifying both in external image-building and in generating internal initiatives to meet the promise of the campaign. The huge success of the Ecomagination campaign (which, incidentally, has more fathers and mothers than there are pieces of the True Cross), spawned a parallel initiative in GE's healthcare business, "Healthymagination," launched in May 2009. The effort was extremely new at the time of writing, but it is predicted that its impact as an internal rallying cry will far outpace its value as a corporate publicity tool.

In light of these systemic issues controlling the environment of company publicity, the tactical model for this communication should have these components: equity and consistency, transparency and collaboration, and internal publicity and recognition.

Equity and Consistency

In order to nourish company publicity in a healthy way, an effort should be made to ensure that the communication group unearths good material and interesting case studies from every division in the company. This can be challenging, especially when those responsible for a promising initiative are unaware of its potential for publicity or, as engineers, their description of it does not make for appealing publicity material. Nonetheless, a sustained effort to prospect for good stories always yields results if handled equitably and consistently over the long term. In time, the process also gets easier as internal constituencies become familiar with the model and undertake initiatives with publicity planned in.

This consistency and equity should apply not only across divisions or business units, but also across geographies.

Transparency and Collaboration

At the same time, it is vitally important that the corporate communication function (including communicators such as the IR team) be aware of the proposed timelines for publicity by various divisions, as well as other corporate communication that may be in the hands of functions such as legal and regulatory. This will minimize, if not prevent, potential conflicts that could cause individual publicity efforts to be less effective than hoped. In some companies, such as PricewaterhouseCoopers, a further effort is made to manage this process by allocating different tiers of importance to various announcements. By exercising some degree of priority management in the way it gives prominence to different publicity efforts, the firm can begin to shape what external and internal audiences to pay attention to.

Transparency also promotes collaboration and shared initiatives because in large organizations it is not uncommon to find independent and unrelated projects across divisions that in fact share common technologies, serve the same market, or exhibit some other features that, if combined, can showcase a company's overall strengths in innovation or customer satisfaction even more compellingly. Even with good knowledge-management systems, a high degree of transparency and collaboration will enable these joint efforts to come to fruition.

Internal Publicity and Recognition

We have described how the principle of "friendly" rivalry can function as an effective stimulant for company publicity that consistently outperforms the mean. Central to this is to ensure that there is equal internal visibility and recognition for initiatives worth publicizing that serve overarching corporate publicity goals. This competition for visibility promotes a full pipeline of fresh material for publicity purposes, but to make this competition possible, it is also important to reward those individuals who make the extra effort to nourish internal communication to promote and facilitate the process. Corporate communicators often lament the difficulty of what is sometimes called "story-mining." Those companies that have instituted some formal recognition for employees who bring forward promising stories and share initiatives across divisions do see a corresponding improvement in the quality and volume of publicizable projects that come bubbling up across the organization.

To achieve a successful, lasting company publicity effort requires a disciplined and coherent approach that treats the entire organization evenly and is to some degree immune to the fads and fetishes of the marketplace. It can be achieved with a tactical model that promotes equity and consistency, and transparency and collaboration, with an appropriate infrastructure of internal communication and reward.

Company Reputation

A well constructed tactical model for company reputation will consist of four elements: an audit of "change appetite"; ongoing monitoring of shifts in the public consensus on issues relevant to the stakeholder audience; a thought leadership program aligned with the change appetite; and a reputation measurement protocol to assess how company reputation is growing or declining.

Just as company publicity can have a profound effect on the ability of a company to realize its operating vision, so company reputation can either block or open the pathways to achieving that vision. Conventional wisdom hews to the view that a good reputation is something that takes years to establish but takes only a moment to destroy. There is certainly more than a little truth to this view, as the terminal exits of Enron and Arthur Andersen have demonstrated. However, we believe that this comfortable formula obscures the true challenges and issues in creating and preserving a good company reputation by suggesting, at least implicitly, that company reputation, once achieved, is something static and established. The reality is in fact quite the contrary.

Reputation is highly dynamic, dependent not only on the continuous flow of actions by the company, but also on the twists and turns of stakeholder opinion, themselves informed by the broad evolution of social mores in all of the societies and polities in which the corporation operates. For us, then, the tactical model that supports the strategic models of advocacy, stewardship, and counsel must be squarely based on an acute understanding of how the shifting sands of social and political change impact the ability of a company to refresh and sustain its corporate reputation.

We can begin by briefly describing what we regard as the cornerstones of corporate reputation. A company with a good reputation usually:

- Sells products and services that are reliable and innovative
- Trains and motivates employees and is regarded as a great place to work
- Offers a good long-term investment for shareholders
- Is a productive partner with government authorities in matters affecting the public interest
- Safeguards and promotes the well-being of communities in which it operates

Within this framework, we need to develop a tactical model that accounts for the dynamic changes in the environment that we have described. The reason for this is that we believe a company's reputation is determined not only by achievements in the categories cited above, but by how well it adapts to changing perceptions in its stakeholders about what constitutes excellence in the categories listed above. For example, one persistent question for companies in recent years has been whether they offer same-sex partner benefits to employees. Before 1990, a company's reputation would have been by and large unaffected by the absence of such an offering. However, at the end of 2008, it could reasonably be argued that not having same-sex partner benefits would be seen as a reputational negative by a significant portion of the U.S. working population.

Choosing Your Place on the Spectrum

Since reputation is a combination of so many different factors, we believe that an important place to start in creating a tactical model for it is to identify an individual company's place on the innovation spectrum. We call it the innovation spectrum because all companies fall along it in their desire, willingness, and ability to change their behaviors and practices. The nomenclature itself is not as important as conducting a realistic corporate self-assessment of the company's "change appetite."

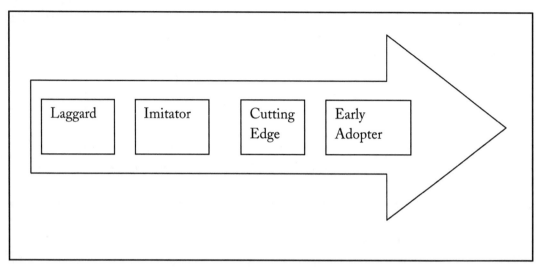

Figure 8.2 Depiction of Company Change Appetite

Among large companies, change in corporate policies and practices is often relatively infrequent, but it is important for any individual company to understand where it falls on the innovation spectrum. For most companies, whether the issues involve transparency, corporate governance, or corporate social responsibility, a position in the middle of the pack best reflects the company's appetite for change in reputational issues. For others with more of a reputation for responding to emerging social trends and business practices, it would be logical for them to be a little closer to the cutting edge. For a few, their corporate reputation is inextricably tied up in the perception that they are evangelists for new corporate behaviors and practices. Their stakeholders expect a fairly high comfort level with change and experimentation. While technology companies were often among the greatest experimenters, particularly during the Internet boom of the late 1990s, it is also often true that consumer products companies find that their stakeholders expect them to be on the cutting edge.

Conversely, there are many companies in regulated industries such as utilities or financial institutions whose reputation is enhanced by the perception that they are slow to change their practices and policies. One unusual recent example of a company choosing an extreme "laggard" positioning is the Post Cereal company, which in 2009 debuted a corporate advertising campaign with the tagline "We Put the No in Innovation."[8] Time will tell whether this positioning works for Post, but it is clearly a conscious choice.

Regardless of the position chosen, what leads to success for an individual company is the extent to which the positioning fulfills two criteria: first, whether the company can credibly adopt the behaviors appropriate to this position, and second, whether its key stakeholders find that positioning desirable and whether it matches their perception of the company. This is another example of the power of the "baseline of expectation" that drives so many aspects of a company's reputation.

Once a company has determined its comfort zone in evolving practices and policies, it can start to carefully track the reputation environment to assess when to consider making changes to move ahead of, alongside, or behind the consensus across a range of reputational issues. We deal at the end of this chapter with the question of measuring reputation, which is a crucial component of the process, but there is one other aspect of creating and sustaining a tactical model for corporate reputation that needs to be addressed. That aspect is thought leadership.

Thought Leadership

Thought leadership has come to mean many things to different people, but we believe in a very simple definition: thought leadership is the demonstration that a company has a point of view about the issues affecting the environment in which it operates and a point of view of its specific market segment. It means that a company is willing to invest in dissemination of these views, not as a form of corporate self-aggrandizement, but in response to the desire from today's stakeholders to understand where the company they work for, invest in, or buy from stands on important issues. The extent to which a company is willing to debate controversial issues or take extreme positions should be dictated by where it has located itself on the spectrum of innovation, but almost all companies today recognize that some form of thought leadership is an essential ingredient of corporate reputation.

Employee Relations

A successful tactical model for employee communications embraces four key ideas: immediacy, comprehensiveness, interactivity, and authenticity.

The much-discussed transition to "knowledge work" over the past 40 years has had a profound impact on the importance of recruiting, retaining, and motivating a company's workforce. (See the discussion of corporate culture in Chapter 5, "Corporate Culture's Increased Significance.") The relative complexity of workplace processes in today's market, the power of retained knowledge about the company's business, and the importance of collaboration have all made the individual worker a more important contributor to the value of a business than at any time in history. This development is reflected in the growth of the employee relations discipline itself but also in the expansion of workplace services for employees and the increased commitment by many companies to corporate social responsibility programs. These programs are certainly designed in part to reflect an increased interest on the part of customers in doing business with an ethically responsible company, but also because workers are attracted to companies with a strong and visible commitment to local and global communities.

This enhanced engagement by employees with the "whole company" they work for gives us the insight to formulate a tactical model for employee relations. If we can caricature the old model as being about health benefits and traditional HR concerns, then the new model could hardly be more different, embracing the widest possible range of topics from innovation programs and environmental initiatives to sensitivity training and team-building. If the old model was focused on communications to a specific worker shift, and geared to an individual plant community, then the new model is based on the idea that any individual employee has a right and an obligation to be well informed about the whole company.

At the same time, we have seen the centralized "command-and-control" approach, which resulted in one-way, "one-to-many" communications being replaced by two-way, "many-to-many" conversations. While consultation and consensus have been described as serving the emotional and psychological needs of the newest generation of workers, the "Millennials," workplace culture throughout most of the North American and European markets has been moving in this direction for many years.

The need to communicate with employees not only about their area of operations but about the whole company, and to do so in a way that fosters a conversation rather than commandments, is amplified by the omnipresence of new communications technologies, leading to an important third tactical principle. This is based on the fact that it is no longer possible or desirable to segment employee communications, walling off information for one group of employees from communications to another group. Today's water cooler extends from Nome to Jakarta, from accounting to the field sales force, and from senior vice president to intern. The new communications technology, abetted by the fact that most companies today are not walled gardens but networks of owned and leased operations, also means that there is for all practical purposes no such thing as a purely internal communication. Any internal communication, especially the sensitive or controversial one, will find its way to the outside world in a matter of minutes.

The convergence of these forces suggests that the most effective tactical model for employee communications should be based on the following principles:

- Employees should regularly receive information about the whole company, not just about their individual unit or about their personal HR needs.
- Internal communications in the age of the horizontal corporate structure should, as far as possible, resemble a discussion rather than a one-way, one-time communication.
- While employees should receive information specifically relevant to them, employee communications should not be compartmentalized, and all employees should receive the same messages about key issues. Furthermore, all internal communications should be drafted as if they will be read by outside parties.

Within the construct created by these principles, a tactical model for employee communications also needs to address the question that bedevils most practitioners: How does one decide which communications channels to use for which types of communications? In the age of online and video communications, this question is often asked in tandem with the related question: How can employees be encouraged to consume the communications they are being exposed to?

The answers to these questions will naturally vary according to the industry sector a company occupies and whether the majority of its employees work in offices, on the factory floor, or on the road. It will also vary by geography, especially as it relates to Internet connectivity. In spite of advances in technology, there are still many parts of the world with no access to the company intranet and no individual email addresses. However, we believe that there are two fundamental ideas that can help answer these related questions.

The first relates to the stewardship aspect of the strategic model. If transparency and authenticity have become hallmarks of the well-respected corporation as a whole, then good stewardship of the corporate brand would also demand that companies create communications models that are respectful of those two attributes. In other words, as an employee I will interpret positively, irrespective of the content, those communications that:

- Reach me first and directly
- Treat me as an adult human being and respect my intelligence
- Are truthful and comprehensive, answering more questions than they pose
- Give me an opportunity for questions and comments
- Tell me clearly how I need to modify how I think or what I do as a result of the communication
- Tell the story using media that make the telling as effective as possible

At first blush these characteristics would appear to be very basic, but experience shows that corporate internal communications frequently fail to meet even these standards. All too often, they are late, either hopelessly terse or frustratingly verbose, or delivered third-hand by messengers who don't have answers to questions and have no way of receiving feedback.

Nonetheless, these standards do provide some guidance for our tactical model of employee communications. They suggest that corporations should look at internal communications methodologies as having, in most cases, the hierarchy of effectiveness shown in Figure 8.3.

Effectiveness	
High	—Face to face, individual meetings
	—Face to face, small group meetings
	—Face to face, town hall meetings
	—Live webcast/video- or teleconference meetings
	—Interactive text-based intranet meetings, with polling
	—Video to the desktop/podcasts
	—Intranet postings
	—Non-digital text/visual materials
	—Email
Low	—Voice mail
Interactivity & Engagement	*High* *Low*

Figure 8.3 Hierarchy of Internal Communication Effectiveness

This ranking of methodologies is, of course, not ironclad, nor are we suggesting that individual approaches are mutually exclusive. In many instances, companies will need to use email first, in order to satisfy the need for immediacy. This is a satisfactory compromise, if the initial communication is followed up by communications with more warmth and credibility. There are also some corporate cultures in which voice mails from the CEO, because of her charisma or employee work styles, might be the most empathetic and effective methodology.

Our underlying opinion, however, is that, where possible, priority should be given to in-person interaction or other communications methodologies that favor interactivity.

With the increased use of online communications, companies are increasingly facing the challenge of non-compliance or under-utilization. In fact, company intranets increasingly resemble the restaurant that baseball aphorist Yogi Berra said was so crowded nobody goes there anymore. Companies begin by posting masses of information to intranet sites, fail to distinguish differing levels of importance, and then don't keep the site fresh. After an initial flurry of activity, employees visit less and less often and then finally don't visit at all. While there are draconian methods for enforcing compliance such as mandating digital signatures, we believe that this method should be used sparingly. Stimulating employee compliance for routine communications is usually effective if the navigation to the information is simple, if the information is relevant to the employee's daily work flow, if it enables employee response, or is entertaining in its own right. This is not always an easy combination to achieve, but with strong writing and a balanced use of video, the company intranet can become a daily rest stop in the employee's day.

At the time of this writing, professional journals and the media are filled with articles discussing the use of micro-blogging in communications, including employee communications. Micro-blogging providers such as Twitter and Yammer offer to host corporate communication networks for messages of not more than 140 characters. Our prediction is that for some companies with cultures that embrace new communications or that have operations that would benefit from brief, frequent communications, internal micro-blogging could become the new instant messaging.

Public Information about the Company

Historically, public information about companies was restricted to professional directories, regulatory filings, historical print and audiovisual materials disseminated by the company itself, or its sales and marketing materials. Other than regulatory filings, the extent of such material was dictated by cost and the interest on the part of management in sharing information. Other than such sales materials, which could be obtained from the company itself, most public information was distributed across numerous text-based storage and retrieval systems such as libraries.

An entire industry of information wholesalers existed to track and archive such material for use in business and by the general public. Gathering information posted in mandatory regulatory filings from different sources was also an enormously arduous and time-consuming task. Of necessity, the amount of company information on all but the very largest corporations was extremely limited. However, since 1999, when the World Wide Web first became available for public commercial use, the amount of public information about companies and other large organizations has exploded. The average corporate website for a big multinational carries more than 100,000 pages of text and video, a number that is hardly likely to shrink in the coming years. Even though companies still distribute large amounts of printed materials, to all intents and purposes their dissemination of public information about themselves has moved to the web. Thus we will not be discussing traditional print materials independently.

The period since the early 1990s has also not been kind to these corporate sites. Except for companies that have re-launched their sites, most corporate websites are an aggregation of many generations of material linked together in often-unintuitive ways. On top of these generational shifts, a decade and a half of mergers and acquisitions, divestitures, and other corporate reformations have resulted in patchwork sites where content and graphic standards, site architecture, and navigational tools vary from business unit to business unit. In sites with ineffective governance structures, portions of the site are using sophisticated video streams and linking to social media, while others are stuck in a pre-Web 2.0 aesthetic, polluted by flash animation and corporate Muzak in inappropriate places.

In addition to these structural problems, the advent of social media has created a new set of challenges relating to the ownership and control of content that begins to appear on corporate websites or on other sites to which the corporate website might be linked. Cisco Systems, for example, has made a concerted effort to provide a social community based on common interests with its stakeholders, including discussion forums and links to other social media such as Facebook.[9] The model they are embodying adheres to many of the precepts of two-way communications but is one that clearly presents some significant challenges. The obstacles to developing an effective tactical model would have to include:

- Content that does not reflect strategic corporate messages
- Navigation and visual identities that are not consistent across the entire corporate site
- The potential loss of control brought about by interactivity

In light of these challenges, how can we create a tactical model that supports the strategic mission to develop strong and positive relationships with stakeholders while advocating for the corporate brand?

The key to tackling this problem is to remember that the issues of governance, organization, and engagement are human and cultural problems, and that no "technical" solution on its own is sufficient to overcome them. We therefore advocate a pragmatic approach that combines technical and human elements to create a solution. The human element requires a concerted look at internal politics in order to achieve the best consensus possible on web integration across geographies and operating units. This requires a high-level discussion among business leaders, not a technical one delegated to web engineers.

The technical element relies on using meta-tagging on an amplified scale to ensure that existing web content and new web content is searchable on a fundamentally greater scale than it is today in most companies. This is because one of the biggest problems with managing corporate information in a large multinational company is that it is still very poorly searchable, both on the website itself and in the source databases from which the information is drawn. A more aggressive tagging strategy not only helps searchability on the site but will also aid in finding strong, recent, and relevant information in company databases. Tagging will also play a key role in search engine optimization, ensuring that stakeholders can get more quickly to the corporate content they are seeking.

The web consultants at Bowen Craggs publish an annual survey of corporate website effectiveness in collaboration with the *Financial Times*. They regard sound "construction" and good governance as two key aspects of effectiveness. They also look at message, contact (how easy it is to contact the company), society (CSR offerings), and how well a site uses web technology (podcasts). Their top rankings, which in 2009 included Roche, BP, Nokia, and Siemens, are reserved for companies that, in their view, do the best overall job with each constituency—

investors, job seekers, customers, and the media—keeping the needs of all these stakeholders in balance.[10]

In today's transparent world, an effective tactical model for a company's public information should (1) find the balance described by Bowen Craggs, (2) exhibit consistent messaging through textual and visual consistency, and (3) exhibit the ease of use made possible by advanced tagging technologies.

Investor Relations

The tactical model for investor relations should:

- Remember that the fundamental nature of the conversation has not changed
- Acknowledge that Web 2.0 has created a continuous rather than intermittent conversation
- Seek out the investor in all of the new media in which he now swims
- Establish a tone that conveys the idea that the investor is part of the management team, with the right to make recommendations and ask deep questions

In many companies, investor relations are frequently still siloed away from corporate communication in a reporting structure tied to the CFO or the Treasurer. This puts the discipline outside the purview of corporate communication. However, since it is central to our thesis that in today's environment it is no longer possible, even if it were desirable, to isolate one stakeholder communication stream from others, we include here some thoughts on a tactical model for investor relations.

From a formal perspective in the practice of investor relations in the past decade, more has changed than in perhaps any other discipline. Anchored in the securities legislation of 1933, the practice of investor relations followed a clear compliance pathway for decades, dictated by the rituals of quarterly and annual reports to the SEC, timely disclosures of material events, and audited financial statements. Whereas most individual investors followed their portfolios by reading earnings reports and heeding the advice of their stockbrokers, large institutional investors benefited from quarterly conference calls with company management and individual meetings and telephone conversations.

By 1999, the SEC had decided that there was a fundamental inequity in the fact that as a result of this distinction, institutional investors had access to market-moving information in advance of the individual investor. In October 2000, the commission adopted Regulation Fair Disclosure (Reg. FD), which mandated that companies share information with both types of investors at the same time. The regulation states that: "Whenever an issuer, or any person acting on its behalf, discloses any material nonpublic information regarding that issuer or its securities to [certain enumerated persons], the issuer shall make public disclosure of that information …simultaneously, in the case of an intentional disclosure; and…promptly, in the case of a non-intentional disclosure."[11] The purpose of the regulation was to even the playing field.

In 2002, following the revelation of spectacular misdeeds at Enron and WorldCom, Congress rushed through the legislation known as Sarbanes-Oxley. While its provisions have been harshly criticized in some quarters for harming the competitive edge of U.S. securities markets, the legislation has also been lauded for strengthening internal financial controls in public companies and restoring confidence in the public markets. It can certainly be argued that the runup in market values between 2003 and 2007 represented a vote of confidence in the system.

However, the global financial meltdown of 2008 has placed every aspect of the financial markets in question, including the practice of investor relations. In particular, the financial uncertainty created by the economic slump has caused many companies to abandon a practice known as "earnings guidance," in which management would give investors their best estimate of revenues and profits for future quarters. In a first-quarter-2009 survey by the National Investor Relations Institute (NIRI) more than one-third of the 600 companies surveyed indicated that they were curtailing or limiting guidance because of difficulties in forecasting cash flow during the credit crunch.[12]

Parallel to the regulatory changes we have described, the evolution of the Internet also created opportunities and exposures in investor communications. These changes revolved principally around the disclosure opportunities represented by the new medium. Traditionally, public companies satisfied their disclosure requirements by issuing press releases through middlemen such as commercial newswires. The SEC, however, pressed corporations to take full advantage of the Internet as a disclosure mechanism, and companies now routinely post their earnings releases and other statutory filings on their websites.

In August 2008, the SEC published updated guidelines for the use of corporate websites and blogs that illustrated how companies could meet their disclosure requirements without issuing traditional press releases. In April 2009, the New York Stock Exchange and NASDAQ moved toward compliance with the SEC guidelines by publishing a proposed rule change outlining the ways in which listed companies could disclose important information without issuing a press release.

Some companies have already adopted a compromise Regulation Fair Disclosure-compliant (Reg. FD) solution. Rather than issue a full press release, they issue a "Notice & Access" release, which is essentially a calling card directing readers to full-text information on their corporate website or on a third-party financial site such as Thompson Reuters. In the meantime, the SEC has pushed the adoption of XBRL (eXtensible Business Reporting Language), a computer language that breaks down the language barriers presented by different reporting formats. On June 1, 2009, the SEC required 500 of the largest U.S. companies to submit financial data using XBRL.[13] The number of companies covered by the rules will grow over time. With XBRL, financial analysts and investors will be able to pull information directly from the SEC or individual companies that can be compared with data from hundreds of other companies in one individual step rather having to laboriously create a spreadsheet by hand from different sources. As the implementation of XBRL widens, the SEC hopes that the volume of data that can be analyzed at little or no cost will more effectively lead investors to both warning signs and opportunities.

Even though changes in investor relations practices tend to be slow, it is probable that over the next few years the majority of public companies will adopt a wide range of activities made possible by Web 2.0. Intel recently became the first major company to offer e-proxies, the ability to vote at annual meetings through the Internet. A number of companies are also pulling questions from investors to use in analyst calls. Google has opened a web forum for investors to post questions to be asked at its annual meeting. European companies such as Bayer offer full IR communications for mobile phones, including SMS services. When eBay's corporate blogger, Richard Brewer-Hay, used Twitter to comment live on his company's analyst call while it was in progress, the investor relations world was electrified.

Some of these practices will undoubtedly fade away after their moment in the sun. Others will become an established part of the investor relations universe. We believe that the tactical model in this volatile arena needs to take account of the fact that the fundamental purpose of investor relations remains unchanged. Its function is to answer these questions: "How does management see the business? What challenges do they have and how do they propose to overcome them?" Skillful and ongoing responses to these questions will still result in the ultimate goal of investor relations practice, which is to create for management the reputation for sustained strategic credibility. At the same time, the transition to Web 2.0 has robbed management of control of the great "set pieces" of the past—the annual report, the quarterly earnings call, the annual shareholders meeting.

Instead of relying on these carefully scripted, intermittent moments, the tactical model of the future needs to be based on a continuous flow of interactions that enlarge the scope of the conversation with investors. To the extent that XBRL gives investors and analysts a tool to dig deeper into the data of thousands of companies within a single query framework, investor relations practitioners should expect to get more probing questions more frequently than today.

The communication channels may grow or shrink in popularity, but the rules of engagement have changed irrevocably.

Notes

1. Jack Beatty, *Age of Betrayal: The Triumph of Money in America, 1865–1900* (New York: Alfred A. Knopf, 2007), 62.
2. Robin Toner, "Washington at Work: The Clintons' Health Care Nemesis: The Man Behind 'Harry and Louise.'" *The New York Times*, April 6, 1994, final ed.
3. Michael Lind, *Up from Conservatism: Why the Right Is Wrong for America* (New York: Simon & Schuster, 1996).
4. Kathleen Fearn-Banks, *Crisis Communications: A Casebook Approach,* 2nd ed. (Mahwah, NJ: Lawrence Erlbaum Associates, 2002), 86ff.
5. Milton Friedman, "The Social Responsibility of a Corporation Is to Increase Its Profits," *The New York Times Sunday Magazine,* September 13, 1970.
6. www.timberland.com.
7. Louis V. Gerstner, Jr., *Who Says Elephants Can't Dance? Inside IBM's Turnaround* (New York: HarperCollins, 2002).
8. "Post Original Shredded Wheat Cereal Puts the No in Innovation," Post Foods LLC Press Release, April 16, 2009.
9. www.cisco.com.
10. www.bowencraggs.com/ftindex.
11. www.sec.gov/answers/regfd.htm.
12. www.niri.org/media/news-releases.
13. www.sec.gov/rules/final/2009.

CHAPTER NINE

Performance

The Measures that Determine the Success of Communication

Metrics are essential to good management, since many executives understand that you cannot manage what you cannot measure. Objective and scientific information, *not* emotional and anecdotal data, provides a context for evaluating performance, provides guidance for making mid-course adjustments, helps to identify competitive weaknesses and obstacles to overcome, and provides the basis for goal-setting. Setting clear research goals and objectives keeps an organization focused. Otherwise the aphorism "If you don't know where you're going, any road will get you there" defines your actions.

Quality research starts with a clear vision of the reason for the research, as well as focused goals. That vision should articulate long-term strategy as well as operational actions and tactics. Consider appropriate and affordable metrics, tools, and techniques, such as media content analysis, Internet assessment, trade show and event measurement, polls and surveys, focus groups, experimental and quasi-experimental designs, ethnographic and corporate culture studies, observation, participation, or role-playing techniques.

Walter K. Lindenmann defines public relations measurement and evaluation as

…any and all research designed to determine the relative effectiveness or value of what is done in public relations. In the short-term, PR measurement and evaluation involves assessing the success or failure of specific PR programs, strategies, activities or tactics by measuring the outputs, outtakes and/or outcomes of those programs against a predetermined set of objectives. In the long-term, PR measurement and evaluation involves assessing the success or failure of much broader PR efforts that have as their aim seeking to improve and enhance the relationships that organizations maintain with key constituents.

Those who supervise or manage an organization's total communications activities are increasingly asking themselves, their staff members, their agencies and consulting firms, and their research suppliers questions such as these:

- Will those public relations and/or advertising efforts that we initiate actually have an effect—that is, "move the needle" in the right direction—and, if so, how can we support and document that from a research perspective?
- Will the communications activities we implement actually change what people know, what they think and feel, and how they actually act?
- What impact—if any—will various public relations, marketing communications, and advertising activities have in changing consumer and opinion-leader awareness, understanding, retention, attitude and behavior levels?[1]

According to Gilfeather and Lindenmann's "Guidelines for Measuring Relationships in Public Relations," successful communication measures fall into four categories: outputs, outtakes, outcomes, and outgrowths.

Ethical Considerations for Research

Organizations both public and private adhere to the 1975 regulations[2] that create an Institutional Review Board, or IRB, which is defined as follows:

> Institutional Review Boards (IRB) are used to ensure the rights and welfare of people participating in clinical trials both before and during their trial participation. IRBs at hospitals and research institutions throughout the country make sure that participants are fully informed and have given their written consent before studies ever begin. IRBs are monitored by the FDA to protect and ensure the safety of participants in medical research.
>
> An IRB must be composed of no less than five experts and lay people with varying backgrounds to ensure a complete and adequate review of activities commonly conducted by research institutions. In addition to possessing the professional competence needed to review specific activities, an IRB must be able to ascertain the acceptability of applications and proposals in terms of institutional commitments and regulations, applicable law, standards of professional conduct and practice, and community attitudes. Therefore, IRBs must be composed of people whose concerns are in relevant areas.[3]

Outputs measure how a company projects its image and identity through quantifiable actions such as the total number of stories, articles, or "placements" that appear in the media; the total number of "impressions"; the number of people who might have had the opportunity to be exposed to the story; and an assessment of what has appeared. A number of services and tools are available to provide metrics through a content analysis of the company's presence in newspapers, TV, radio, Internet, blogs, and other social media. Outputs can also assess thought leadership positioning of the company by tracking executive speeches, editorials, appearances, events, sponsorships, and press conferences. Often a company program, event, or meeting is designed to achieve exposure and create awareness for an organization, its products, or its services. Metrics assess and analyze the people who attend, as well as any results of the event such as press interviews and stories reported. And surveys intended to measure changes in opinion or attitude often determine if target groups have received the intended information or concepts.

The idea of Advertising Value Equivalency (AVE) has been around for many years. It has generated much debate in the public relations industry, that debate focusing on both its reliability and validity. Many people are attracted to it because it appears able to put a dollar value on media coverage and, by extension, allows media relations people to compare their results with advertising. Yet the measure has a number of problems, and it is important to be aware of both its strengths and its weaknesses.

The Commission for PR Measurement & Evaluation does not endorse Ad Value Equivalency (AVE) as a measurement tool, but one hears all the time from people who have bosses or clients who demand it.[4]

Outtakes are a common and important way to measure the impact of an action or campaign. Outtakes determine if a target group received the message in the first place, if they paid attention to the message, if they understood or comprehended the message, and if they retained and can recall the message.

Measurement tools and techniques can include surveys, focus groups, polls, and corporate culture (ethnographic) studies. Corporate communication, as advocate for the company, strives to inform and persuade groups concerning the issues that it considers important. To do this, companies use awareness and comprehension metrics, as well as recall and retention metrics.

Outcomes measure the effectiveness of a particular action. They assess what the communication has created, for example: leads for future business; consideration of products and services; sales; an increase in stock price; and changes in awareness, attitudes, and behavior. After all, outcomes are concerned with positive attitude, opinion, and behavior changes that result from company actions.

Research techniques include quantitative surveys (in-person, by telephone, by mail, by fax, by email, through the Internet, in malls, etc.); focus groups; qualitative depth attitude surveys of elite audience groups; pre-test/post-test studies (e.g., before-and-after polls); ethnographic studies (relying on observation, participation, and/or role-playing techniques); experimental and quasi-experimental research projects; multi-variate studies that rely on advanced statistical applications such as correlation and regression analyses, Q-sorts, and factor and cluster analysis studies.

Outgrowths measure the impact of all corporate communication actions on the positioning of an organization in the minds of its constituents. The focus of many corporations is the impact of communication on the organization's reputation among all constituencies or stakeholders. Executives now know that reputation risk can outweigh legal risk. That is, the cost of losing reputation can dwarf the financial losses of a lawsuit. They also know that corporate communication actions have a positive impact on Return on Investment (ROI) and generate revenue, reduce cost, and often avoid the costs associated with a crisis. Corporate communication executives also contribute to the corporation's overall value creation, the increased worth of the enterprise.

Measuring Reputation

Reputation is an increasingly valuable corporate asset. Ron Alsop has stated the corporate reputation management issue succinctly: "The first step in managing reputation is measuring it. You can't manage what you can't measure. Companies must mine the research data and learn how they are perceived by different audiences and which factors drive their reputations."[5]

When asked in the CCI Corporate Communication Practices and Trends Studies to rank-order 13 management functions, corporate communication executives have consistently ranked "Manager of company's reputation" and "Counsel to the CEO and the corporation" as the phrases that best describe their primary function as chief communication officer.[6] Table 9.1 indicates the percent of the respondents who ranked the functions as #1.[7]

The challenge for corporate executives is which tool to use, since consensus in measuring cor-

TABLE 9.1 Primary Role of the Corporate Communication Executive in Companies

Counsel to the CEO & the Corporation (*new-2002*) [16.7% #2]	23.3%
Manager of company's reputation [14.8% #2]	18.0%
Source of public information about the company [14.5% #2]	12.9%
Driver of company publicity	10.2%
Manager of the company's image	9.7%
Advocate or "engineer of public opinion"	6.8%
Manager of relationships –co. & NON-customer constituencies	5.8%
Branding & brand perception steward (*new-2002*)	5.0%
Member of the strategic planning leadership team (new-2009)	3.5%
Manager of employee relations (internal comm.) (*new-2002*)	3.4%
Manager of relationships --co. & ALL key constituencies	1.9%
Support for marketing & sales	1.8%
Corporate philanthropy (citizenship) champion (*new-2002*)	1.8%

Source: CCI Corporate Communication International Studies

porate reputation remains elusive. In addition to the longest-running measure, the annual *Fortune* "Most Admired" ranking, there are numerous other measures of reputation. Most of them are proprietary and include the Walker Information Corporate Reputation Report (proprietary), Y&R Brand Asset Valuator (proprietary), Reputation Institute "RQ Gold" (proprietary), ORC Reputation (proprietary), NOP World (formerly Roper), Landor Image Power (proprietary), and the PriceWaterhouseCoopers Reputation Assurance Framework (processes). Add to these CoreBrand (Brand Power and Brand Equity rankings), Rating Research (letter grades similar to Moody's Investor Services), and Delahaye Medialink Worldwide (reputation index based on print and broadcast items).

Other rankings of corporate reputation include *The Financial Times*'s World's Most

Respected Companies; Millennium Poll (Prince of Wales Business Leader Forum and The Conference Board); *Business Ethics*'s 100 Best Corporate Citizens, and *The Wall Street Journal* Reputation Quotient (Harris Interactive).

As long as these measures use a variety of corporate attributes, differing methodologies, and survey different respondents, corporations will continue to appear at different positions in the rankings, depending on the survey selected.

Ron Alsop's *The 18 Immutable Laws of Corporate Reputation* offers practical advice on these three phases of reputation management. For establishing a good reputation, he recommends that a company focus on what it does best, its "most powerful asset," delivering packages on time in the case of FedEx. He suggests that companies measure their reputation by using some standard such as *Fortune*'s America's Most Admired; CoreBrand; or *The Wall Street Journal.* "Learn to play to many audiences," says Alsop. That is, focus company messages on the needs of individual constituents. Reaffirm, reinforce, and revise company values and ethics. (Alsop cites Johnson & Johnson as an example.) Being a model by practicing corporate citizenship, like McDonald's, helps to create a positive reputation. Convey a sustainable, adaptable, and compelling corporate vision. For example, L.L. Bean founder Leon Leonwood Bean's Golden Rule is "Sell good merchandise at a reasonable profit, treat your customers like human beings, and they will always come back for more."

Alsop also suggests that companies create emotional appeal, as do Johnson & Johnson, Harley-Davidson, United Parcel Service, Coca-Cola, and Whole Foods Market.

To protect a good reputation, a company needs to recognize its shortcomings through continuous evaluation and self-improvement initiatives. A corollary to this is to stay "vigilant to ever-present perils," such as the scandal at *The New York Times* over reporter Jayson Blair filing non-factual stories as news in 2003. Companies with positive reputations, according to Alsop, make their employees reputation champions. And because misinformation can spread instantaneously, he cautions corporations to "control the internet before it controls you," to "speak with a single voice" and avoid mixed messages, and to "beware the dangers of reputation rub-off," for example Martha Stewart's trial for insider trading and her relationship with Kmart, or Ben & Jerry's alternative corporate culture and its acquisition by global giant Unilever.

And when a company's reputation is damaged, Alsop offers practical ways to repair it. In cases such as J&J and Tylenol in 1982, Coca-Cola in Belgium in 1999, and Martha Stewart in 2001, he suggests that these companies managed crises with finesse. When there is a crisis, the company must fix it right the first time. He cautions, however, never to underestimate the public's cynicism, citing the long public memory over the gay community's boycott of Coors Beer, Philip Morris changing its tobacco-associated name to Altria, and the Alaska oil spill from ExxonMobil's tanker the *Exxon Valdez.* He notes that "being defensive is offensive," as in Ford's Bridgestone/Firestone tire recall, Enron's fraud scandal, and WorldCom's "cooking the books."

Finally, "if all else fails," Alsop instructs, "change your name." It worked for ValuJet in 1996, when it became AirTran after a tragic crash. Philip Morris became Altria. Andersen Consulting became Accenture before Enron.

Alsop's "Name That Company"

Can you match the current and former company names?

Current Company Name	Former Name of the Company
1. Navistar International	a. Consolidated Foods
2. Unisys	b. International Harvester
3. Nike	c. NGC
4. Sara Lee	d. Sandoz and Ciba-Geigy
5. Clarica	e. Sperry and Burroughs
6. Target	f. Blue Ribbon Sports
7. Diageo	g. Bell Atlantic and GTE
8. Dynegy	h. Dayton Hudson
9. Verizon	i. Mutual Life of Canada
10. Novartis	j. Guinness and GrandMet

Source: 18 Immutable Laws of Corporate Reputation
Answers: 1-b, 2-e, 3-f, 4-a, 5-i, 6-h, 7-j, 8-c, 9-g, 10-d

OXFORD METRICA REPUTATION MEASUREMENT DESCRIPTION

In recent years, there has been a renewed focus on moving beyond opinion-based measures of reputation to identify specific reputation-related activities that have a cause-and-effect impact on shareholder value and growth in revenues and profitability. One such effort by the firm Communications Consulting Worldwide uses econometric modeling and multivariate statistical analysis to establish the relationship between communications and the business outcomes that matter most to a company: stock price, revenue, employee retention, and market share. This approach is called "Value Measurement."

Another new candidate in this field, the analysis of changes in share price and market capitalization to assess the impact of reputation, is the "Reputation Equity Management" system (REM) developed by Porter Novelli in collaboration with Oxford Metrica. By looking at a range of so-called "value events" for an individual company and comparing them to similar events for industry peers and a 10-year database of 1,000 publicly traded companies, the REM analysis can identify the reputation drivers for that company and measure the growth or decline in its reputation equity both chronologically and against a customized set of peer companies. Using its range of proprietary algorithms, Oxford Metrica's analysis is able to strip away overall market and industry impacts on a company's market value. What is left behind is the company's reputation equity, a measure of the sustained strategic credibility that is compelling to investors.

Oxford Metrica has also produced research (see Figure 9.1) showing that even a year after a crisis, it is possible to discern significant differences in market value between what are called "recoverers" and "non-recoverers" that are clearly attributable to the perception of how each group handled the crisis that befell it.

Management consultants, professors, and public affairs executives now consider 30–50 percent of a company's value to be in intangible assets, such as these components of corporate equity: familiarity with the company, overall impression of the company, perceptions of the company, and likelihood to support the company. Corporate reputation encompasses all of these intangibles.

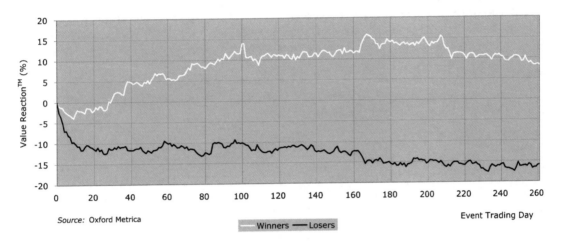

Source: Oxford Metrica

Figure 9.1 Recovery from Reputation Crises: Winners and Losers

CEOs overwhelmingly believe that reputation is important to their companies, that a good reputation helps sell products and services, that a good reputation helps attract employees, and that a good reputation increases credibility in times of crisis.

According to one CEO, "If you're not managing your corporate reputation, you're wasting a global corporate asset." Another observed, "A reputation is an incredible asset, one you can't appreciate until you lose it. And when you do, every aspect of business gets harder and more costly."

"If you lose dollars for the firm by bad decisions, I will be understanding," instructed Warren Buffett. But "if you lose reputation for the firm, I will be ruthless."

Companies with positive reputations are

- 7 times more likely to have consumers buy products/services at a premium price
- 5 times more likely to have their stock recommended
- 4 times more likely to be recommended as a good place to work
- 3 times more likely to be recommended as a good joint venture partner
- 1.5 times more likely to receive the benefit of the doubt

Corporate communication is the appropriate function for managing reputation because it influences the intangibles, communicates with all stakeholders, and strengthens strategic relationships. Corporate executives who measure, manage, and communicate their financial and non-financial performance not only contribute to the increased valuation of their firm, but they also improve their firm's attractiveness for long-term investors and for new investment capital.

Notes

1. Walter K. Lindenmann, "Guidelines for Measuring Relationships in Public Relations," www.instituteforpr.org.
2. U.S. Federal Register (45 CFR Part 46) (1975).
3. See the IRB Operations and Clinical Requirements list provided by the FDA's Office of Health Affairs. This document is intended to help IRBs carry out their responsibility to protect research subjects. Also see the March 13, 1975, Federal Register, and the Technical Amendments concerning "Protection of Human Subjects" (45 CFR Part 46).
4. Bruce Jeffries-Fox, "Advertising Value Equivalency," www.instituteforpr.org.
5. Ron Alsop, *The 18 Immutable Laws of Corporate Reputation: Creating, Protecting, and Restoring Your Most Valuable Asset* (New York: Free Press, 2004), 25.
6. CCI Corporate Communication Practices and Trends Studies, www.corporatecomm.org/studies.html.
7. www.corporatecomm.org/pdf/CCI200StudyUSOctober 2009.pdf.

PART FOUR

Strategic Adaptation for Global Practice

In Part 3, we traced the evolution of corporate communication as a response to the social and political pressures facing corporations in the late-19th and 20th centuries as they struggled to justify and defend the enormous power and influence they had begun to wield over individuals, communities, and even countries. Our review of this evolution set the stage for a discussion of the strategic models available for communication specialists of today as advocates for the corporate point of view, as stewards of the company reputation, and as counselors to the leadership of the business. These strategic models in turn provided a framework for the tactical models to be used in fulfillment of the roles predicated upon the three strategic models. We concluded Part 3 with an overview of the increasingly sophisticated tools being used to measure corporate reputation and the tools designed to nourish and sustain it.

In Part 4, we face the challenging task of divining the way forward for corporate communication. In order to do this we are required to determine which of the forces of change we described in the preceding chapters are cyclical, that is, determined by transient economic, social, and political forces, and which are secular, more fundamental changes to the purpose and practice of corporate communication transforming them for the long term. In thinking through these issues, we are mindful of the adage that change never acts as quickly as we expect, but when it does, its influence is more profound than we could ever have expected.

To assist us in this task we rely not only on an analysis of the impact on corporate communication of the global financial crisis of 2008–2009, but also on research conducted in 2009 by the CCI Corporate Communication International for its annual Corporate Communication Practices and Trends survey. In addition to the statistical data provided by the survey, we conducted lengthy interviews with corporate communication leaders of a number of established multinationals to ask them how they see the interplay of cyclical and secular trends influencing the role and practice of their discipline. We have incorporated in this part verbatim commentary from seven of the most insightful practitioners in corporate communication to underscore and illuminate the trends that we see guiding the way forward.

CHAPTER TEN

Corporate Communication

The Way Forward

Our discussion of corporate communication began with an assessment of the ways in which three powerful forces were transforming the principles and practices governing the relationship between the corporation and its stakeholders:

- *Globalization*—a quantitative shift in the globalization of the world economy that has created a qualitative change in how businesses need to communicate;
- *Web 2.0*—a transformation in the adoption, use, and consumption of information technology;
- *Corporate Business Model: The Networked Enterprise*—an evolution in the nature and purpose of the public corporation that both influences and is influenced by the other two forces at work.

The first of these, Globalization, was the emergence of a truly integrated global marketplace in which the business supply chain operates not only around the world but in a single time zone—now. Complementing and enabling this integration was the rise of emerging markets, not only those markets in the first tier—Brazil, Russia, India, and China—but also the next 11: Bangladesh, Egypt, Indonesia, Iran, Mexico, Nigeria, Pakistan, The Philippines, South Korea, Turkey, and Vietnam.

The second major force was the transition of the Internet toward Web 2.0, in which massively greater bandwidth has facilitated a wide range of new communications media from blogs and podcasts to YouTube and Twitter. We explored how this technology transition has created a newly connected, networked, and transparent web of social media that has had a profound impact on the speed and extent to which stakeholders can interact with the corporation and with each other in ways that have both positive and negative implications for the corporation.

Both of these major transformations have taken place against the backdrop of a long-running decline in the reputation of corporations and other institutions such as religion, the justice system, organized sports, and the media. In the developed economies, this skepticism regarding corporations has been given additional force by the emergence of the Millennial Generation, whose lack of faith in and loyalty to the corporation were so expertly conveyed in Ron Alsop's *The Trophy Kids Grow Up*.

These long-term trends were hijacked by the global financial crisis of 2008 and the economic turmoil thus created, which forced a transformation in fundamental business models, beginning with the evaporation of almost an entire industry sector, investment banking. Manufacturing and the auto industry teetered on the edge as Congress debated their fate, and media companies faced an uncertain future. The Tribune Company, to name just one media entity, filed for bankruptcy at the end of 2008, and many other media enterprises are struggling to create a viable future.

At the same time, the convergence of an increasing number of sovereign countries onto similar economic growth models that characterized the period through 2007 came to an abrupt halt. The global crisis revealed that there were huge disparities between the economic stability of different countries. The prosperity of Iceland, Ireland, and parts of Central Europe was savagely compromised by the economic downdraft, and other economic disparities both within and between countries has created new social and geopolitical tensions. Even in countries less severely impacted, government measures taken to respond to the crisis have fundamentally transformed, perhaps for a generation, the relationship between private enterprise and civil society. According to the *Financial Times* (December 10, 2009,) Jeffrey Immelt, CEO of General Electric in a December 2009 speech at the U.S. Military Academy at West Point declared, "We are at the end of a difficult generation of business leadership ... tough-mindedness, a good trait, was replaced by meanness and greed, both terrible traits...Rewards became perverted. The richest people made the most mistakes with the least accountability...[we should welcome government as] a catalyst for leadership and change."

The uproar over CEO pay and the use of corporate jets represents merely the leading edge of this transformation that has led to more government "ownership" in banks, insurance companies, and automobile manufacturers, necessitating a new kind of collaboration among government, corporations, and labor.

The CCI Corporate Communication Practices and Trends Study 2009[1] included in-depth interviews with the corporate communication officers who chose to participate. One of the open-ended questions asked, "What are the **top three critical issues** in corporate communication today?"

The responses of the CCO's echo the themes of this book. One executive at a global financial services corporation said the issues are: "how to use communication as a tool to help achieve business goals, how to achieve high performance with restricted resources, and how to stay ahead of today's militant populism."

An executive at a global pharmaceutical corporation answered, "building trust internally and externally, reputation management, transparency," and in the context of the U.S. debate on health care reform, "positioning the CEO." In addition to rebuilding trust between front-line workers and management, another executive added engaging and addressing multiple stakeholders on issues and plans in a world where there are multiple venues for dialogue and in which anyone can be an "expert, as well as demonstrating leadership in a complex and rapidly changing world."

Others noted the need to build credibility and trust to mitigate reputational risk—assessing, protecting, and measuring reputation. Adding strategic value and demonstrating the function's value in driving business performance is another concern. Another issue is the ability to develop important relationships with investors, media, and non-governmental organizations (NGOs), and investing the time to focus on relationship building.

For another executive the issues are "Responding to a changing workplace; relating to a changing employee body. Younger people are coming in and expecting social media. In corporate communication it is authenticity; keeping the creditability of your firm high; operating in an authentic, transparent manner. Keeping employees motivated and keeping your constituents happy in today's economy." For another, "skepticism of what we're saying and the sincerity of the message is a major concern."

One executive indicated the concern we have addressed in this book for strategic adaptation, saying there is "a yearning for the path" and noting that there "is a refusal to acknowledge when things are changing. And for another the issues are a changing media landscape, turning employees into advocates, and the need for authenticity.

Paradoxically, the first decade of the 21st century also witnessed an unparalleled increase in the social demands being placed on the corporation. In less than a decade, the expectation that corporations would place corporate social responsibility at the core of their activities has moved from the periphery to the middle of the opinion spectrum. Sustainability, which grew out of the concept of environmental stewardship, came to embrace social justice and a wide variety of micro- and macroeconomic issues including "fair trade" and community building. To a greater or lesser extent, corporations have responded positively to the concept that ending poverty and healing the planet belong in their job description alongside providing jobs, inventing new products and services, and offering shareholders a return on their investment. In Chapters 2 through 9, we described how corporations have and should respond to these challenges at both a strategic and tactical level.

It is too early to predict confidently which of these transformational shifts will have the greatest long-term impact, but what is clear is that the consequences of human exploitation of natural resources have not been abolished merely because economic development has slowed or reversed itself. If anything, the consequences of climate change and industrial production have made it more important than ever that we find collective solutions to the challenges of hunger, access to clean water and breathable air, sustainable practices in energy production, and construction and packaging, to name just a few.

Recent events have revealed ways in which certain aspects of globalization may no longer be sustainable. As we mentioned in Chapter 1, catching salmon in Alaska, shipping the fish to China for filleting, and then back to stores and restaurants in New York is ethically questionable in an energy-constrained environment. We also noted that, in the spring of 2008, when gasoline prices rose above $4.00 a gallon in the United States, California strawberry farmers bought land in New England to short-circuit transportation costs to their East Coast markets.

Constrained economic circumstances and the impact of global warming will unquestionably influence many other resource-allocation decisions in increasingly complex ways.

The direction of the global economy is likely to be murky at best for some time to come. What we can predict is that the combination of the long-running secular trends we have described in this book, combined with the most perilous economic environment since the 1930s, will create conditions of unprecedented volatility for global corporations. Public opinion and political opinion in different parts of the world regarding how corporations should behave is likely to shift more quickly than at any time in recent history. From one day to the next, public expectations on subjects such as corporate governance, social justice, environmental stewardship, or data privacy could turn 180 degrees. Sovereign governments, struggling to find a solution to intractable economic conditions, may seek convenient scapegoats in the form of large corporations, tying their hands through new regulations while at the same time blaming them for failing to create more jobs. Economic nationalism, whether in the form of "Buy China" or "Buy America," will further complicate the conditions in which corporations will need to operate. Finally, the hyper-transparency made possible by Web 2.0 and demanded by an anxious public will ratchet up public scrutiny to a new level.

In this context, the successful management of corporate communication will require all of the skills and capabilities that we outlined in Chapter 2, as well as the most finely tuned judgment, wisdom, understanding, and integrity that the corporate communication professional can muster.

The CCI Corporate Communication Practices and Trends Study 2009 cited above asked the corporate communication officers another open-ended question, "What top three **trends** in corporate communication do you see in companies?"

Business Performance. For a financial services executive there is "more emphasis on business goals, less emphasis on social priorities, and doing more with less." The down market forces companies to become more efficient and demands that "Fortune 500 companies operate as lean entrepreneurs." One executive revealed that his company had "five people worldwide" and that lean organization at the top meant that he provides counsel for groups to "beg, borrow, and steal" from one another to collaborate on any communication.

Social Media. One executive interpreted the number of invitations to seminars on social media and its growth as an indication of the importance of that trend. Another executive sees social media as the number-one issue, observing that the challenge is "how to best manage it and how to better engage this type of media. My point of view is that you can't manage it but rather how do you best engage that particular type of media." The fascination with social media as a communication tool, according to another executive, creates "a reinvestment in communication." It also creates the challenge to "understand how independent bloggers are affecting the messages that companies are trying to send out, and how can companies adapt to this environment from a reputation and crisis communication standpoint." As the current media relations practices—traditional press and news releases—disappear, companies adapt to the new media and are "focusing on measurement and social media." One executive observes in this trend "the rapid pursuit of social media—some of which may prove to have little ROI." For an executive in financial services, it is "developing content that can be leveraged on both internal and external platforms. I'm trying to avoid using the term 'social media.' Employees now have Twitter and Facebook accounts. It's important to develop content that can be used both ways."

Employee Communication. The emerging importance of employee communication and engagement is another trend that is "more about building networks and connections among internal and external stakeholders and less about producing stuff," according to one executive. Another observed that "HR is not in charge of internal communication, but rather a partner with corporate communication, so the trend is to 'pull in HR as partners.'" Another executive has "an employee website that we update and add to five times a day to our employees and constituents. We try to duplicate the communication experience that people have in their own home. Video's, not like the old days."

Authenticity and Transparency. The trend toward "more authenticity and transparency requires a deep commitment to ethics and values." Another trend is "face-2-face communication; making leadership approachable—which translates into how executives should respond at town hall meetings." And for external communication, the issue is also "Authenticity, transparency, & responsiveness: it's never going to be good enough in this 24/7 news cycle whether it's internal or external. 'No comment' isn't going to fly in this new world. If you don't tell your story someone else will tell it for you on CNBC. It's all about speed."

Convergence of Marketing and Public Relations. Several corporate communication executives see "more integration between corporate communications and marketing especially in using unpaid media and social media to build reputation." That is resulting in a "graying of lines between PR and marketing." One manufacturing executive is "disturbed by this trend" and "concerned about corporate communication being rolled up under the chief marketing officer," noting that corporate communication "needs to report to the CEO or to the COO." There is also "integration of messaging and positioning from branding to financial communication, to employee communication, to customer communication." Such convergence of communication functions has taken place at IBM and at Johnson & Johnson.

Business Acumen and Alignment. For another executive, alignment is a key issue. It requires corporate communication executives to be business professionals first, and to understand the business. However, it is not necessary to be a CPA or MBA to have "business acumen, to understand how to make money and how to drive revenue, how to sit with a business unit to give them counsel and earn credibility."

Sustainability. "More and more companies are looking at the environment and sustainability and how do they best communicate that."

The Way Forward

In response to the forces we have discussed, we believe that corporations will need to explore a variety of pathways that take into account the changing role of the corporation in society, as well as the changes in communications technology we have outlined. Depending on the industry in which they operate, some companies will focus more intently on the tactical question of dealing with the new social media, while others will predominantly wrestle with the strategic issue of changing relationships with stakeholders. A few companies will successfully synthesize the opportunities presented by both the tactical and strategic challenges and will arrive at a fresh way of nourishing and developing the corporate brand. In both cases, a hair-trigger sensitivity to changing mores, combined with a strong backbone, will be key requirements.

Additionally, the CCI Corporate Communication Practices and Trends Study 2009 asked the corporate communication officers, "What do you see ahead for the practice of corporate communication?"

Strategic Advisors. "What may change is greater reliance on corporate communication to be strategic counselors and advisors on business as well as communication decisions, and to have our finger on the pulse of what's happening in the industry, with customers, with employees and in the community—and to advise accordingly. There will be less reliance on corporate communication to report on things for employees, and more on engaging employees in getting involved in what is happening. The staffs may be smaller but will be relied on more for their expertise."

Business Acumen. "I think the function continues to be perceived as critical and its importance will sustain in the future, but it will increasingly be expected to make a tangible contribution to business success." "In the past having a department with good writers might have sufficed, but that's no longer good enough. We're going to find people who will have to be more focused on business. I think that model is going to shift more to *true* business people that will be good at communication."

Increased Stress. One executive notes that his "blackberry is always on and gets emails in a round of golf" on his day off. In such an environment, "people will burn out; tremendous pressure; stress levels have become difficult."

Trust and Authenticity. "Demonstrating authenticity and rebuilding trust in organizations are critical in a world where cynicism prevails. Corporate communicators are in a unique position to potentially rebuild trust in our institutions and organizations by making sincere connections and meaningful networks, supported by open, honest communication and dialogue."

New Ways to Communicate. "I think the traditional approach to corporate communication will fade away as companies grapple with multi-generational workforces and a variety of stakeholders. Add to that mix the splintering of mainstream media and the onset of social media, and the departments that can truly morph into business problem-solvers will survive while others who rely on traditional press releases and employee newsletters will eventually be found irrelevant."

Changing Nature of Communication. "The ability to communicate with each other has changed. More and more people are becoming citizen journalists. We need to understand it and then get our hands on it. The idea that you can suppress knowledge or keep something secret is gone. The idea that employees can only vocalize their frustration internally is gone."

Building Relationships. "People have a huge reliance on technology. They communicate with reporters via email and in some way we've lost some of the most fundamental part of what we should be doing which is trying to manage relationships....It still comes back to personal connections and driving certain types of thoughts and behavior....At the end of the day it's not so much this new media...it's about how do we really connect to people in ways that make relationships richer and stronger over time."

Jack Bergen, communication leader at Alcoa, who early in his career accompanied Jack Welch of General Electric during their town hall discussions, sees a contemporary tactical adaptation in using mediated "Speak Up" intranet sites. These enable staff and employees to participate in video and webcast interviews and Q&A forums with the CEO, CFO, and other corporate leaders. The intent, aside from disseminating information and gathering employee attitudes and ideas, is that dialogue not only builds relationships and company culture but also contributes to more professional, productive, and profitable communication.[2]

For Harvey Greisman, formerly at Mastercard, the challenge is to integrate outreach to influential bloggers covering the company and the industry and to develop an appropriate relationship with this new audience that meets their instantaneous and persistent needs.[3]

Catherine Mathis of Standard & Poors, the former SVP of Corporate Communications at *The New York Times* company, sees particular challenges for media companies seeking to sustain their reputations in a supremely difficult time for the media industry. Mathis challenges her professionals to keep up to date with Twitter, search engine optimization, and social media releases. Meeting the needs of shareholders, journalists, readers, and advertisers poses a special challenge for a media company, particularly when the competition reports on your business. For example, Mathis got a phone call at 5:00 P.M. on Christmas Eve 2008 from Russell Adams, a *Wall Street Journal* reporter, saying that he heard that *The New York Times* company was exploring the sale of the Boston Red Sox, which had yet to be announced by the *Times* company. The response was not to comment on M&A rumors. So *The Wall Street Journal* posted its story on its website and in print the next day, scooping *The New York Times*.[4]

Dick Martin, who has written three books (*Tough Calls, Rebuilding Brand America*, and *Secrets of the Marketing Masters*) for the American Management Association since his retirement as EVP of Corporate Communication at AT&T, sees the disappearance of the "bright" dividing lines that used to exist in corporations among functions such as corporate communication, marketing, and government relations. For him, the disappearance of these boundaries betokens a profound shift in the role and responsibility of the communication function away from targeted messaging to discrete audiences and toward a more holistic view of how society as a whole looks at companies. He attributes this change to the decision by companies such as Microsoft, IBM, and GE to place in the chief marketing role someone with a public relations and corporate communication background. Martin sees this as the CEO's bet that generating revenue now requires a broader view of the environment and customer needs beyond the traditional Ps (product, placement, price, promotion) of marketing. Walmart's reputational ills are a good example of its rethinking of the relationship with its core values, its customers, and its value proposition as people increasingly see a direct relationship—a holistic relationship—between the company and its products and services.[5]

Jon Iwata, in a speech on the future of the profession, observed:

> Some believe that the integration of marketing, communications and CSR is not only logical, it is inevitable—because of all the changes in the *external* environment—the need to speak with one voice across advertising, sales promotion, events, websites, the media, analysts, bloggers and the like. Of course there is value in message consistency, especially now, with the diffusion of media. But you don't need to go through all the work of structural integration to achieve that kind of alignment. (In fact, as many of you will attest, integrating organizations often impedes unity, at least at first.) Rather, I believe the most powerful advantage of putting these teams together is that we have combined our culture with our brand, and our values are the foundation of both.

Experts in the workplace and experts in the marketplace are now on one team.

As never before, people care about the corporation behind the soft drink, the bank account, the computer. They do not separate their opinions about the company from their opinions of that company's products and services—or its stock, for that matter.

People care about the behavior and compensation of the company's executives…how the product was produced, and by whom…how the company treats its employees and suppliers…how it impacts the environment. Now, maybe people always cared about these things—but, really, how much could they know about what was happening inside our companies? Today they have an unprecedented view into the corporation's actual behavior and actual performance.[6]

Ray Jordan of Johnson & Johnson believes that the convergence of new media with a shift in the public's expectations about corporations has placed an entirely new kind of dialogue at the core of corporate communication. He contrasts traditional communications, which conventionally revolve around a known information endpoint, with the more fluid and comprehensive conversation companies are now having with their stakeholders. Using terminology popularized by Google, he believes that the public conversation with stakeholders, whether via Twitter, YouTube, blogs, or wikis, better resembles a "perpetual beta" release than an exchange of fixed opinions. From this perspective, the role of the corporate communicator is to manage the flow of this uninterrupted conversation rather than present stakeholders with "completed" information products. He concedes that few, if any, companies have learned how to do this successfully. His response has been to make selective investments in social media, to learn by doing. He believes the benefits will outweigh the risks.[7]

Reid Walker, former Global Corporate Communications Officer at Lenovo, believes that his company's "rootless" legacy gives it a perspective that other corporations could emulate. He believes that in the current volatile environment, having operations on every continent but no true headquarters gives Lenovo a unique ability to think in a borderless way. Without a head office, country, or culture of origin, he says that employees in any country feel accountable for the good reputation of the enterprise. To cope with the volatility of the present moment, he meets monthly with his "Reputation Council," comprised of leaders from each business unit, to assess emerging threats and move quickly to take advantage of opportunities.[8]

Tom Buckmaster, VP, Corporate Communications at Honeywell, uses a four-stage communication model for increasing the business value within a multinational organization: service provider, business resource, business partner, value creator. "As a service provider, corporate communications executes client requests and provides great service measured by volume and quality in developing messages, managing events, writing, and editing."

"As a business resource, our focus is on clients' business objectives and is measured by the importance of the projects and the function's impact on them. We address underlying business problems, attend planning meetings, and educate clients on corporate capabilities. Baseline business acumen is required of the corporate communication staff."

"As a business partner, we solve business problems collaboratively and are measured by the impact we have on problems and the value created for the partner. As a value creator, we deconstruct business problems with the partner, engage clients, as well as offer critical and analytical thinking, to solve the problem as a result of a deep understanding of the client's value chain. We aspire to always move up this value chain…maturing our contribution and expanding our impact and relevance as we do."[9]

For these executives, in short, corporate communication as a strategic management function creates measureable value for the corporation. Each of these leaders provides insight to support a variety of approaches that can be tailored to the needs of individual organizations.

As George Friedman says, "At a certain level, when it comes to the future, the only thing one can be sure of is that common sense will be wrong. There is no magic twenty-year cycle; there is no simplistic force governing this pattern. It is simply that the things that appear to be so permanent and dominant at any given moment in history can change with stunning rapidity."[10]

Thus corporations and individuals capable of the continuous adaptation we have discussed throughout this book are prepared to meet such uncertainty. In fact, they expect and welcome it, since it fits both tactical and strategic assumptions about how organizations can behave successfully.

Notes

1. CCI Corporate Communication Practices and Trends Study 2009, www.corporatecomm.org/studies.html.
2. Interview with Jack Bergen.
3. Interview with Harvey Greisman.
4. Interview with Catherine Mathis.
5. Interview with Dick Martin.
6. Jon Iwata, "Toward a New Profession: Brand, Constituency and Eminence on the Global Commons," 2009 Distinguished Lecture, Institute for Public Relations, Yale Club, New York City, November 4, 2009.
7. Interview with Ray Jordan.
8. Interview with Reid Walker.
9. Interview with Tom Buckmaster.
10. George Friedman, *The Next 100 Years: A Forecast for the 21st Century* (New York: Doubleday, 2009), 3.

PART FIVE

Guidelines

Introduction

Parts 1 through 4 of this work have provided a theoretical framework for understanding the forces that influence corporate communication. We have explored the strategies necessary for the chief communication officer at a large, multinational corporation to meet the challenges of the transforming global economy, the revolution fueled by digital media, and a substantially changed business model in the 21st century. We have analyzed the forces and the communication challenges that they represent for the global corporation. In examining these forces and how they are interwoven, we have presented strategies for the corporate communication professional and business leaders to help them deploy effective communication as a strategic business asset in today's global economy.

In Part 5 we provide concrete and specific guidelines, tactics for how to organize and execute effective communication for the contemporary practitioner. Based on our research (see Chapter 1, Table 1.1, "Core Functional and Budget Responsibilities of Corporate Communication"), we identified eight functions that, taken together, form the core functions of contemporary corporate communication practice—Media Relations, Public Relations, Communication Strategy, Crisis Communication, Communication Policy, Executive Communication, Reputation Management, and Employee Communication. A strategic framework with tactical guidelines for effective communication offers a powerful and valuable combination for effective practice.

In Part 5, more than a dozen "Guidelines" offer leading practices for putting corporate communication strategy into action. This part also provides caveats for practice by suggesting situations that might require actions that seem counter-intuitive—especially guidelines for when to close the rule-book and use experience, expertise, and judgment. In light of frequent regulatory changes in the areas of investor relations and transparency and disclosure, we recommend that readers consult www.sec.gov for U.S. government updates to these rules.

These Guidelines serve as a foundation for individual and organizational adaptation:

A. Corporate Communication Strategy and Policy
B. Crisis Communication
C. Media Relations
D. Employee Relations
E. Global Relations
F. Corporate Citizenship and Table of Organizations
G. Core Competencies for Corporate Communication
H. Investor Relations and Sustainability
I. Transparency and Disclosure
J. Reputation Management
K. Transaction Communication
L. Affiliate Relations
M. Thought Leadership and Executive Relationship Management

Guideline A: Corporate Communication Strategy and Policy

In classical management theory, responsibility for a particular activity lies with the functional unit best suited to executing this activity. Financial management is handled by the finance function, legal work by the office of the general counsel, and employee benefits by human resources. While information technology has changed almost everything about how these functions carry out their responsibilities, not very much has changed about their scope, and—for the most part—each of these functions has a clear reporting line into the office of the chief executive.

This cannot be said for the corporate communication function. Neither the scope of this function nor its reporting line within organizations is subject to a standard model, and within large and small corporations today, there still remains a wide spectrum of practice. Depending on the history and culture of individual companies and the industry in which they operate, we can find corporate communication as a component of many different functions:

- The marketing organization: Corporate communication here is seen primarily through the lens of corporate public relations, marketing the image of the company.
- The human resources team: In this structure, corporate communication tends to be focused on internal communications and community relations.
- The legal department: In this model, all communication is seen as potentially risky and needs to be governed by risk-management thinking.
- The chief financial officer: This is a less-common approach but is seen in companies who think of corporate communication primarily as investor communications.

In many companies, yet another factor comes into play. Because the uncertainty in the minds of many management teams about the proper role and function of corporate communication, responsibility for this activity is simply given to the company officer trusted as having the "safest pair of hands." In this model, corporate communication is someone's extra job that rarely leads to value creation. Not surprisingly, this mode of practice tends to lead to greater structural volatility for the communication function as it shifts from one pair of hands to another.

Finally, there are organizations in which the chief corporate communication officer reports directly to the chief executive, a model for which communication professionals have argued strenuously over many decades. Indeed, when executed in a truly strategic manner, this approach offers the best of all worlds in terms of corporate reputation, giving the necessary access to the communication team to advise on reputation risk and obtain senior management backing for reputation-building activities. However, this model by itself does not guarantee that the function will be given the opportunity to maximize its effectiveness. Even though proximity to the ultimate decision-maker, the chief executive, generally confers power and responsibility, this model sometimes simply creates a gilded cage in which the chief communication officer becomes the chief courtier of the CEO. This kind of setup rarely leads to a communication function with independent and measurable strategic objectives.

Inside the Tent

Parallel to the question of where corporate communication is located within an organization is the question of which communication activities it controls. Almost all corporate communication teams are responsible for corporate media relations, now including the blogosphere, but usually manage some of the other main communication functions:

- Employee communications
- Investor relations
- Government relations/public affairs
- Community relations
- Crisis communications
- Corporate social responsibility
- Corporate philanthropy
- Corporate image and identity

Much ink has been spilled and many corporate turf wars fought over the question of which of these functions should report to corporate communication and which should not. These struggles miss the point and obstruct a serious discussion of the key question: What is the purpose of corporate communication, and how should the effectiveness of the function in executing the purpose be measured? As the themes of this book suggest, we believe that the core purpose of the corporate communication function is to enhance and protect the reputation of an organization with all of its key stakeholders. In our discussion of measurement (see Chapter 9, "Performance: The Measures That Determine the Success of Communication"), we suggested some ways in which corporate reputation can and should be measured. Whatever the metrics chosen, these should be tracked year over year, and the chief communication officer should be made accountable for improvements or declines in these measures.

The implication of choosing the enhancement and protection of corporate reputation as the core purpose is to suggest that it is less important which functions corporate communication controls and more important that its views are incorporated into the way in which these functions are executed. The corporate communication leadership should play an integrative role to ensure that all of the communication activities of the enterprise are aligned and mutually supportive of the goal to enhance and protect its reputation. In order to achieve this, the corporate communication group must have wide access to and extensive consultation with all of the communication streams cited above. Only in this manner can the communication strategies and policies of an organization be internally consistent.

The Emperor's New Clothes

It is not just the explicit communication streams, however, that should lie within the purview of the corporate communication team. If the core purpose of the function is to enhance and protect reputation, then it is essential that it also be evaluating all of the key activities of the organization on a continuous basis to check for potential reputation risks and reputation-building opportunities. With the imprimatur of the office of the CEO, the communication leadership should have free rein to ask key reputation questions. The following are a few hypothetical examples:

- Are we comfortable with the privacy implications of our customer data retention practices?
- Do our new employee orientation procedures promote a strong sense of our corporate culture?
- Are we building strong relationships with the non-governmental organizations (NGOs) focused on our industry?
- Do our environmental practices lag or lead?
- Have we thought through the broader implications of canceling a sports sponsorship?
- Does the compensation structure for our sales force create incentives to engage in dubious practices?
- Has the age and lack of diversity of our board of directors become a potential liability?
- Should we be more engaged in industry panels on technology innovation?

What this short list of typical questions demonstrates is that the corporate communication function is the only place in the organization in which two things can happen: (1) commercial decisions of every kind are subject to a critical test of their reputation impact, and (2) they are examined to determine whether these decisions can build reputation along with their primary purpose. This is not a role for the faint-hearted, and we are not suggesting that the CCO become the authority on all the decisions made in the company. What we are suggesting is that, like the child in the fairy tale, the corporate communication officer is the only person with the job to point out that the emperor—corporate leadership—is wearing no clothes in the parade.

Guideline B: Crisis Communication

Crisis communication defines that area of communications that comes into play when an organization's reputation, as well as its human, physical, financial, and intellectual assets, comes under threat. This threat can be sudden and abrupt, as with an airline accident, or remain dormant for a long time before bursting into public view. The latter form of crisis might be a long pattern of employee harassment by a manager that finally comes to light, or a technical incident of financial fraud that is revealed only as the result of an investigation into an unrelated matter. Regardless of the cause, each crisis requires rapid and consistent decision-making and communications with all affected stakeholders, under conditions of stress in which fundamental ethical issues are often in question. As a result of this interplay of factors, organizations can sometimes exacerbate the reputational impact of a crisis through poor communications. From time to time, communications themselves can cause a crisis. Gerald Rattner, the Managing Director of a leading British retail jewelry chain, famously called his own products "crap" in a speech to the Institute of Directors, leading to a significant destruction of shareholder value. Crises, whether self-inflicted or a product of independent events such as an earthquake, are so crucial to organizational reputation because it is in the speed and effectiveness of the response that good management and leadership is demonstrated—or, conversely, found wanting.

Issues Mapping and Crisis Communications Planning

While many crises appear to come out of nowhere, in most instances missed warning signals could have alerted the organization to an impending problem. This is why one of the first principles of effective crisis management is sound planning to identify risks and put in place protective plans to manage a crisis, should one occur. This process needs to take two distinct but equally important pathways. In the issues-mapping process, the planning team looks at every discrete activity of the organization and maps all of these activities against evolving trends in the global environment. The purpose of this planning exercise is to identify hidden threats and vulnerabilities that, if ignored, have the potential to create a crisis. Some of these vulnerabilities exist in plain sight. A coal mine obviously needs to have highly developed safety and materials handling procedures and a process for responding to any breach of mine protocol. A pizza company, however, might not have thought through the ramifications of its on-time delivery guarantee in a sales model that relies on independent delivery contractors. In one well-known example, a national pizza company created incentives for its drivers to drive too fast (delivery guarantee) while at the same time imposing the costs of maintaining the vehicles on the drivers themselves (independent contractors). The gradual and then accelerating up-tick in accidents involving pizza deliveries could have been predicted by an analysis of the company's basic business model.

The issues-mapping process creates a robust picture of the threats and vulnerabilities faced by the organization. Once this process is complete, crisis communication planners can determine which of them require changing the way the organization behaves and which require careful monitoring and messaging. The issues identified should form the basis of a keyword search-monitoring system to help give the organization early warning of media interest in these topics. Traditionally, these monitoring systems have focused on print and television media, but today, organizations should also be monitoring websites, search engines, blogs, and podcasts, as well as video content on social networking communities such as YouTube and MySpace.

At the same time, the planning team should also be conducting a crisis communications planning exercise. This consists of identifying the individuals in the organization who would need to participate in any crisis response and the ways to quickly convene them in the event of a crisis. Once the individuals have been identified, a communications and decision-making structure appropriate to the organization needs to be created in order to provide an effective crisis response. In most crises, rapid information-gathering and decision-making can be the deciding factor that separates success from failure. Another key step at this stage of the process is to pre-position company data and messaging in all of the communications channels available. This could include background materials explaining the company's value chain, as well as websites hidden from view that can be activated at a moment's notice.

The final step in crisis communication preparedness should be a drill that puts the system through its paces. A good drill will help accustom the participants to their roles and unearth hidden limitations in the crisis communication plan. At one German refinery, the crisis communication drill revealed the fact that in an actual emergency all of the phone lines in and out of the facility were shut down except the line to police, fire, and safety centers, posing a significant problem for communicators trying to deal with other stakeholders in a crisis.

Crisis Onset

Even with effective preparedness and a careful issues-mapping process, a crisis can take hold with lightning speed and come from any direction. Sometimes the first news an organization has of a crisis brewing is a call from a reporter. In other situations, the crisis is already being covered on every TV channel and across the web before management has become aware of it. Regardless of where the bulletin originates, the crisis response team should be called into action immediately. In the early minutes and hours of the crisis, this team needs to develop the clearest possible view of how the crisis is likely to unfold. The team needs to answer some key questions quickly. These include:

- What steps still need to be taken to manage the crisis itself?
- Who has been harmed, and what actions will make them feel the organization has responded effectively?
- Who should speak on behalf of the organization? How frequently and in what forum?
- Which third-party entities—academics, think tanks—can be enlisted to speak up in defense of the organization?

These questions are designed to help steer the organization toward a successful and credible public perception in the midst of a crisis. In effective crisis communication, the key determinants are not the severity of the crisis itself, but how the crisis is being handled and how well the company involved is communicating about it. This means that in the early hours and days of a crisis, even before the facts are clear, companies that understand crisis response set up a system by which they can deliver regular updates to the media and other stakeholders. Even if there is little fresh information to share, the mere fact of offering to communicate has beneficial effects: it expresses confidence that the organization can handle the crisis and implicitly accepts the proposition that "the public has a right to know" what's going on.

At the same time, companies need to be aware of "dribbling out" new information. If one of the key goals of any crisis communication effort is to get the story off the front page, then releasing small amounts of new information simply has the effect of providing the media with fresh details to cover for another news cycle. This does not mean the organization under attack shouldn't be communicating. In fact, one of the most valuable tasks to be performed by the crisis team is to identify the impact of the crisis on various shareholders and prepare messaging that is specific to each stakeholder. In the wake, for example, of a spectacular business loss, the media will be simultaneously pursuing stories about the impact on investors, employees, the communities in which the company has major operations, the philanthropic organizations the company has supported, and so on. Being ahead of the game by having reached out to these stakeholders and having appropriate messaging sends a strong signal that the management team understands the ramifications of what has happened and is not just thinking about themselves and the fate of the company but is sensitive to the entire community. One frequently neglected stakeholder group in a crisis is employees. Sharing unfolding events with internal audiences in a timely fashion is one of the hallmarks of good crisis communication.

Mid-Crisis Response

Each crisis has its own rhythm, but there is often a moment of stasis in mid-crisis when an organization takes or doesn't take the steps that will be required to ensure long-term recovery of its reputation. This is the phase in the crisis when the immediate danger is usually over, when the broad outlines of the crisis and what might have caused it are clear, and there has been wide coverage of the events involved in the crisis. Having survived the intense heat of the acute phase of a crisis, decision-makers often mistakenly believe that the worst is over. This moment, however, is precisely the moment at which recriminations begin, and the public—regulators, the media, and analysts, depending on the crisis—are looking for answers. Astute crisis managers understand that this is the moment when decisions with long-term consequences need to be made. Some of the questions that come up in this phase are:

- Do we need to retain an independent panel of experts to assess the origins of this crisis and recommend solutions?
- Do we need to close this business unit or fire the CEO?
- Do we need to fundamentally reshape our relationship with our customers?
- Do we need to recall the product?

History is replete with examples of companies that failed to act aggressively at this stage of the crisis. Consider, for instance, the consumer products company that decided to stop shipping its products but didn't announce a recall. What happened? Retailers simply removed the products from the shelves themselves. Or the company that announced that the CEO would retire, but not for three months. The CEO was gone within two weeks.

At the other end of the spectrum are companies that act too precipitously in their anxiety to find a scapegoat and fire mid-level or lower-level employees before the dust has even settled. This is rarely an effective move, because it short-circuits the period in which the media and the public are still deciding whether there was a fundamental defect in the company's systems that needs to be repaired. Such actions usually evoke negative reactions even when, objectively, a firing might have been justified.

Crisis Recovery

Once the crisis is over, there are still enormous reputational dangers that lurk in the waters. Even when the news is off the front page, the media and the public will still be looking to see whether the organization has fully understood the nature of the crisis and, as appropriate, changed its behavior to reflect that understanding. This usually involves changing company policy or increasing training programs, replacing aging equipment, or agreeing to independent monitoring of its activities for some period of time. To the extent that management changes have been part of the solution, the new management still needs to demonstrate that the problems that fostered the crisis have been resolved, and periodically remind stakeholders that changes have been made. Sometimes this process can take several years. One well-known casual dining chain was the target of dozens of lawsuits charging systemic racism in the workplace tolerated by management. After the crisis blew up, the leadership of the organization made long-term pledges to ensure that minority employees and diners would be treated equitably. The company continued to report on its progress in this effort for many years, winning plaudits from the community for its efforts and its perseverance.

Conclusion

Even in well-run organizations, issues bubble below the surface and can erupt at any time. With a careful crisis-response system in place, however, and a commitment to monitoring emerging threats, most organizations can come through a crisis with their reputations intact and sometimes even enhanced. The key ingredients for good crisis communication are essentially no different from good communication in any setting: openness, clarity, and putting oneself in the other person's shoes, whether that person is an employee, a customer, a shareholder, or a member of the community. The challenge in a crisis is that our natural instinct is to do the opposite—control information, circle the wagons, protect our own, attack the enemy. Overcoming these built-in reactions and looking to the long-term outcomes for the organization are the hallmarks of a true crisis communicator.

Guideline C: Media Relations

As we have discussed throughout this book, the media model has changed dramatically. So, too, the model for media relations have changed and become one of relationship building and reputation management. In practice, these tasks can be accomplished through the traditional method of meeting the press (see Table C1, Meeting the "Traditional" Press), or a range of company actions.

For example, the company can use *free media* to deliver its messages through high-credibility outlets that cover events, speeches, or press conferences. A company *war room* can be responsible for monitoring and responding to news as a way of refuting critics and delivering messages in the instantaneous news cycle. *Paid media*, such as television, print ads, brochures, websites, and mailings can deliver controlled and targeted messages. *Networking* with organizations, unions, politicians, media, and other stakeholders can be used to listen to and develop relationships with a broad set of audiences. *Corporate citizenship* activities reinforce a company's message through its social and environmental contributions and actions. *Operations* such as L.L. Bean's customer satisfaction policy and Starbucks's fair-trade coffee policy underscore company messages and mitigate reputational risk through core business activities. *Partnerships* with NGOs, such as Shell's with Greenpeace, show collaboration on mutually shared industry issues. *Surrogates* such as prominent individuals with high credibility reinforce strategic company messages. And *grassroots* initiatives, such as blogs, websites, bumper stickers, and town hall meetings enlist supporters on behalf of the company.

Table C.1 Meeting the "Traditional" Press

SUGGESTED ACTION	RATIONALE
Be Prepared	In an information society such as ours, having accurate data and timely statistics is expected. Not only are you giving your valuable time to discuss issues and events with the press, but their time is valuable also. So, do your homework, and prepare wisely for a press interview.
Make Your Points	Have three main points you wish to get across. Just as you would in an executive summary of a report, or in a marketing communication, identify clearly the main ideas that make up the message you want communicated.
Be Concise	But avoid yes, no. Show awareness of the space and time limitations of the media by presenting positions clearly and concisely. Although brevity is a virtue, the press also looks for interest. Yes and no answers make the story difficult to write, and uninteresting for TV or radio.
Get Comfortable	Remember that movements and eye contact communicate non-verbally. When meeting face-to-face with the media, prepare for the discussion by making sure that you will not be interrupted. A conference room set aside for outside guests is a good idea.
Tell the Truth	Building credibility with the media begins with their perception of you as a source of accurate and truthful information. Integrity is a valuable attribute. People react positively to people they perceive to be genuine. Being yourself is linked with telling the truth and is part of building corporate integrity.
Use the Printed Word	Prepare for press encounters with a printed statement or press release. The document helps reporters get facts straight—figures, statistics, the spelling and titles of people mentioned. Remember the reporter's job is to report the facts, and getting accurate information includes often complex and detailed data.
Keep Your Composure	The media must attract readers and viewers to sell advertising. Such pressure translates into the search for unusual or controversial angles. Journalists call this the "hook," the means to capture the audience's attention. Offbeat, even offensive questions are a common tactic to elicit an emotional reaction that would make a good headline. So be cool under pressure.
Think of the Reader or Viewer	Remember the importance of the audience in any communication. Consider how remarks would appear on the front page of *The New York Times* or *The Wall Street Journal*, or the local TV news, or on the national news.
Say You Don't Know	When asked a question that stumps you, or requires information or data you do not have at hand, say you don't know. Follow up immediately with plans to get the information, and an offer to contact the reporter later, preferably before their deadline.
Hypothetical Questions, Third-hand Information	Reporters may ask questions that lead to speculation. Such questions are particularly common when corporate officers are asked to comment on possible mergers, anticipated layoffs, or restructuring. Also, if reporters cannot identify the source of their information, politely decline to discuss rumor and hearsay because of your company press policy.
Sensitivity to Deadlines	The daily production of newspapers and TV or radio broadcasts places strenuous demands on reporters to file their stories on time. It is a common courtesy to ask at the beginning of an interview when the reporter's deadline is.
Accessibility	Give reporters a contact number, an email address, and a fax number to indicate that you will be available for follow-up questions as the story is being written, and later as a source for other stories.
Forget "Off the Record."	If you don't want something to appear in print, or broadcast, then don't say it.
"No Comment."	Finally, the press universally interprets the response "no comment" as a ploy to hide something. Say clearly the company does not discuss proprietary issues, or matters that have personal impact on employees.

Guideline D: Employee Relations

The goal of positive employee relations is to promote employee retention, increase productivity and work quality, stimulate innovation, and help attract high-quality employees. In spite of the importance of these goals in increasing shareholder value, employee relations are often an under-appreciated and under-resourced aspect of organizational communications. One reason for this is that relations with and communications to employees reside in so many different places in an organization and are concerned with so many different discrete topics. These include human resources, operations management, training and development, plant management, security, and environment and health, to name just a few. In addition, especially in a large-scale organization, it can be very difficult to communicate efficiently so that the entire organization receives information directly from the source (as opposed to the grapevine, which universally operates more effectively than any official medium).

In order to confront these challenges, we believe that high-quality employee relations depends on four core principles:

- Integration
- Sequencing
- Content and Medium
- Feedback

By observing these four principles, organizations of any size can create effective employee relations.

Integration

A global shipping company was trying to identify why its employees from middle management on down seemed almost impervious to company communications. When feedback was solicited, little came in, and when deadlines for completion of certain activities were requested, they were seldom met. Research into this issue revealed a simple and obvious answer to the problem, which—as in many other large organizations—was obscured by departmental silos. In this company, almost every meaningful departmental function required middle managers and employees to be responsive in October and November of each year. Unintentionally, all of these communication streams had been bunched together. October was the month that re-enrollment in the health care plan began. October was when budget estimates for the coming year had to be locked down. October was the month that performance reviews had to be done so that promotions and raises could be approved. If all these communications streams were not enough, October was also the single-busiest commercial month of the year for customers of this global shipping company. It was no surprise that the company's employee relations system was broken. What was surprising was that it functioned at all.

This example illustrates our first key principle of good employee communications, which is the integration of all the communications going out to employees so that they can be appropriately sequenced. This requires a careful audit of routine communications coming from different departments throughout the year in order to optimize the effectiveness of each stream. It also means setting up a mechanism in the corporate communication function that can monitor and, to the extent possible, influence ad hoc communications to prevent dissonance between different messages. It is important to avoid inadvertent clashes between the announcement of a luxurious new research facility in Ireland with news of job cuts in Kiev. The winner of the 2008 innovation award should not be announced the day before the plant where she works is closed.

This same alertness to the integration of information going to employees applies equally to coordination between internal and external announcements. It is especially important to recognize that employees are never just employees. They are also shareholders, members of the plant community, and parents of children in the local schools, and any internal or external communication needs to be scrutinized for the potential to create tension between these roles.

Sequencing

Dana Corporation is a 105-year-old automotive components company headquartered in Toledo, Ohio. One of the tenets of its employee relations culture is "Tell Dana people first." Making sure that employees hear important news from management directly rather than from other sources is a key principle of good employee communications more honored in the breach than the observance, but it is critical to achieving buy-in and support from employees. Even when securities law dictates that material information be shared widely and simultaneously with the market, best practices in employee relations require immediate follow-up with broad-scale employee communications.

The second critical aspect of sequencing, especially with major announcements, is that employees need to hear the news from immediate supervisors as well as senior management. A well-executed "roll-down" plan incorporates in-person briefings soon after an announcement. Employees should have an opportunity to speak live and directly to their own team leaders to get vital information about how the news affects the employee personally, and not just the broader commercial ramifications for the company.

In a multinational company, time zones also play a critical role in sequencing. Employees in New York who log onto their email before starting their morning commute should not be getting messages about a major corporate announcement from colleagues in Asia before their own morning briefing at headquarters. In order to avoid this problem, companies need to carefully think through when and how they start the sequence. Sometimes this requires overnight email notification, which sacrifices directness. However, if the email is followed up as quickly as possible with in-person meetings and the sequence is explained, employees will accept this necessity.

Content and Medium

The late Speaker of the House Tip O'Neill often referred to the advice he had received from his father that "all politics is local." The same wisdom can be applied to the practice of employee relations. All communications with employees should lead off with and be focused heavily on those issues with specific relevance to the individual employee. In fact, the more significant the information being imparted, the more detailed it should be in providing guidance in exactly how it impacts an employee's daily life and work.

Following universal rules of good communication—clarity, simplicity, brevity—is particularly important in relations with this critical audience, because even in the best-run companies, employees are naturally skeptical and critical of management. This is why ambiguity is one of the most toxic components in employee communications. In order to fight this natural tendency, all communications to employees should be subject to these two tests:

- Anything that can be misunderstood will be misunderstood.
- Given a range of possible interpretations of a statement or event, employees will select the most negative for the company and themselves.

This is especially true when communications to employees are the end-product of a confidential management task force. Managers frequently overlook the fact that even though they have been wrestling with a given set of issues for months, the employee is hearing about or reading about a proposed policy change for the first time. It is all too frequent to see employee communications that, in effect, start in the middle or even near the end of the story. It is crucial to begin at the beginning with the "what" and the "why" before embarking on the details of the "how." In fact, for complex topics that will have a significant impact on employees' lives, it is important to repeat communications and to frequently revisit the subject to probe for understanding and support in the early weeks. Where possible, it is also helpful to involve a member of the target employee group in the communications planning on a confidential basis in order to test for the credibility of the message. Just as a CEO will have his investor relations team prepare Q&A material for the quarterly earnings call, any major employee communications effort should be supported by a detailed rehearsal of the questions likely to be asked. Any hint that management has not thought through obvious questions raised by a new initiative can cause that initiative to die at birth through lack of employee engagement.

The communications channels suitable for employee relations will vary from organization to organization. However, it is important to select communications channels that satisfy the two main goals of any communication: speed and understanding, both rational and emotional. This means that some combination of face-to-face meetings and recorded digital communication is usually optimal. However, research has shown that some forms of digital communication are more effective than others. With all its advantages of speed, email, for example, is not universally available, and in many countries individual company email addresses are still a rarity. In other countries, the volume of email received by the average employee has risen to such an extent that it has been rendered almost useless in communicating anything more complex than brief, neutral information. Company intranets have grown increasingly popular as places to post information for employees. This can be an effective communication tool, but its success is subject to a high degree of variability, depending on the company culture and the quality of the intranet interface. In this information economy, furthermore, all communications have to compete for attention with the blizzard of inputs employees receive each day. This means that taking creative approaches is often required to achieve positive results, and no communications medium should be ignored in attempting to break through the clutter. Videos, CDs, posters, balloons, mouse pads, and rubber gloves have all been used in successful employee communications. However, communicators should never forget that humor is the least universal language of all. Any use of humor in employee communications should be vetted at local levels throughout an organization. In the final analysis, the Golden Rule should be applied to employee relations more than with any other stakeholder, because the future of any organization lies mostly in the hands of employees. Managers should choose a lexicon and communications channels that best convey respect and trust for the individual employee.

Feedback

In this era of social media, it is easier than ever before to obtain employee feedback, a crucial component of good employee relations. Whereas in an earlier age leading companies would conduct annual or biennial surveys of employee attitudes and opinions, almost every communications platform available today comes with a polling widget, making it cheap and extraordinarily easy to obtain employee feedback about company policies in general or about specific issues. Indeed, workplaces occupied by an increasing number of employees comfortable with social media are experiencing an explosion of solicited and unsolicited feedback. However, it is important to remember that the old rules of employee relations have not been abolished:

- Don't ask for feedback unless you are willing to make changes based on it.
- Don't publish feedback selectively or in a sugar-coated manner.
- By and large, employees tell management what they think management wants to hear.

Within the context of these time-honored principles, the new social media platforms offer organizations an astonishing variety of ways to solicit and receive employee feedback. Enterprise social media platforms, corporate wikis, internal CEO Twitter feeds, private video channels, prediction markets, and employee blogs all create novel opportunities for organizations to hear from employees on a continuous basis. Every company will need to create its own blend of these tools. Those cultures that are most comfortable with employees who voice opinions, and who can be tolerant of the occasional mishap, will undoubtedly gain the most from these new communication channels.

Guideline E: Global Relations

In the first decade of the 21st century, a more truly global marketplace was created than had existed at any time since the decade preceding World War I. Notwithstanding some reversals of this trend created by the global recession that began in 2008, the global marketplace created through the emergence of countries such as China, India, Brazil, and Russia has transformed the nature of global relations for multinational companies. There have been two principal drivers for this transformation. The first is the removal of regulatory barriers controlling foreign ownership of business assets in countries such as India, and a parallel reduction in subsidies or protections for home-grown "industry champions." The second is the development of truly global supply chains involving wholly owned and wholly outsourced operations.

The combined power of these two shifts has brought about the decline of global corporate infrastructures based on having autonomous country or regional business units in favor of globally matrixed organizations. In these matrixed structures, responsibility for managing a brand globally, for example, could be headquartered in one country, with transportation and logistics in another. In this model, employees responsible for marketing or transportation would report both to the global manager of their function as well as a country manager in their own nation. Multiple reporting relationships become even more complicated in some contemporary organizations, where an individual could be accountable to a country managing director, a global function leader, a key client relationship manager, and the captain of an ad hoc continuous-improvement task force. All of these developments have profound implications for the practice of global relations, creating some new—and reinforcing some old—obstacles to effective communication.

Trouble in the Matrix

Major global corporate communication initiatives that have the explicit eyes of senior management on them are served well by the matrix. In these cases, significant energy and resources are devoted to ensuring that the initiative is globally aligned and that results are obtained. However, in the day-to-day activities required to nourish the global corporate brand, that energy is sometimes lacking. Under these circumstances, the divergent priorities of global brand managers and individual country leadership can sabotage global corporate brand-building efforts.

The best treatment for this malady is to create an explicit conversation about the problem with key global and country decision-makers. This conversation can then lead to the development of transparent metrics for the "reputational health" of the global brand, independent of country-based financial metrics. These metrics should include overall visibility and awareness scores, as well as track the presence or absence of key corporate messages.

An ounce of prevention can also be worth more than a pound of cure. In order to prevent the misalignment of country and global-brand goals, it is important to involve local communication resources in an individual country at the beginning of the process rather than waiting until the localization phase. Not only will early engagement bring a greater diversity to the planning process, it will also ensure that there truly is ownership of the program by all parties around the world and not just those at corporate headquarters.

The Marketing Silo

The emergence of global corporate and product brand-marketing organizations has significantly strengthened the coordination of brand messages across geographies. The speed with which new initiatives can be launched simultaneously in many regions through such organizations has been impressive. However, rapid regulatory changes in the 21st century, as well as increasing public scrutiny on ethical and other grounds, have meant that compliance and regulatory questions now lie at the center rather than at the periphery of corporate and product brand-marketing efforts. To cope with this evolution, sophisticated corporate and brand-marketing organizations are breaking down the walls of the marketing silo in order to better integrate with other functions. One way they are doing this is by creating permanent cross-functional global teams that are organized on the basis of specific issues such as privacy, the environment, or appropriate marketing to children. By creating these teams, companies can do a better job of avoiding reputational risks produced by initiatives that contravene legal guidelines or community standards of behavior in different countries around the world.

The Silent Scream

The emergence of major new markets in the developing world and the very different population profiles of those countries compared to the developed world have put new pressure on the global/local customization issue. A corporate brand strategy that can accommodate Brazil with a median age of 28 and Italy with a median age of 41 clearly needs to be very carefully articulated. At the same time, the increasing self-confidence and self-esteem of developing world populations makes them more intolerant of communication programs that have not been appropriately customized to them. Unfortunately, differing business and national cultures make open dialogue about these strains extremely difficult, leading to passive resistance and even silent sabotage that can derail communication efforts. Effective corporate communication networks have found that the most successful response to this challenge is to tackle it head-on, creating an explicit conversation about cultural assumptions and differences in the use of ostensibly common terminologies. This process will not eliminate conflicts between the global center and the network, but, as far as possible, it produces a rational, rather than emotional, debate.

Corporate/Brand Dissonance

Globalization has been reshaping the multinational organization for decades. This process was significantly accelerated in the last decade by the proliferation of private equity deals in which the rapid turnover of corporate assets was a necessary response to the debt levels required to finance these deals. The result was an increasing dissonance between a company's product brand portfolio and its corporate brand identity. At the same time, an increased interest in corporate social responsibility and ethical behavior created a more intense focus on corporate reputation and the links between the product and corporate brands. In response to these pressures, corporations became more transparent about the relationship between corporate and product brands, which, in turn, highlighted those instances in which these were in tension. Unilever, for example, has been widely praised for Dove soap's "Campaign for Real Beauty," which embraces female self-esteem. But it has had to deal with criticism that its advertising for Axe personal care products, which is aimed at teenage boys, demeans women.

These tensions can never be completely eliminated, but successful corporations are trying to minimize them by paying closer attention to how the corporate brand is aligned to product brands and acting more nimbly in making brand adjustments across their global networks.

The Three Monkeys—hear no evil, see no evil, speak no evil

One of the biggest challenges in global relations is ensuring that local issues in one country don't bubble below the surface for weeks or months before suddenly blowing up onto the global stage. Every CCO's nightmare is a problem in one country that could have been managed and mitigated at an early stage but remained undetected at the corporate level. The causes of such a situation are not hard to imagine. Bad news is seldom shared in a timely manner in large organizations, and the sheer volume of communication in organizations of this size can make it hard to ensure an effective "signal-to-noise" ratio.

The best deterrent for this problem is to set up a system that incorporates standard diagnostic criteria and that, within broad guidelines, determines:

- which issues remain purely local
- which issues can be handled locally but should be disclosed to the global corporate communication team
- which issues need to be disclosed and handled together by local and global communicators
- which issues are predominantly the domain of global leadership

Some companies use a "traffic light" model to illustrate these principles, but the key component is making sure that local country communicators and managers know what kinds of issues need to be escalated to the global team and when.

In addition to this notification system, we also recommend that companies develop a standardized "issues management" protocol for describing and tracking potential vulnerabilities. By creating a company-wide reporting standard and ensuring that country-specific issues are shared with the global communication team, multinational organizations increase the likelihood that they will catch major reputation problems early. By seeing these reports, the global team can also identify issues common to numerous countries that could suggest a fundamental problem with company practices around the globe. In the optimal scenario, the company can then take steps to address the issue and change practices if necessary before the media or, in the worst case, regulators in multiple countries start to take action. This best practice has become especially important since the Internet reduced the cost of sharing information between different country chapters of global NGOs to practically zero.

Productive global relations in today's marketplace rely on an effective use of web-based tools to manage, measure, and integrate information from multiple countries at the center. This demands having explicit conversations about cultural and linguistic barriers to reputation-building efforts around the globe. It also requires an understanding of the extent to which, in country after country, the line between the corporate and product brand has disappeared and that a healthy alignment of the two must be an ongoing effort in order to enhance and protect the corporate reputation.

Guideline F: Corporate Citizenship and Table of Organizations

Corporations now find that financial reporting alone no longer satisfies the needs of shareholders, customers, communities, and other stakeholders for information about overall organizational performance. Many corporations now use triple-bottom-line reporting, which is also called citizenship reporting, social reporting, or sustainability reporting. They can have one report or several reports that publicly disclose the corporation's economic, environmental, and social performance.

Corporate citizenship has evolved from the actions of philanthropic foundations, other nonprofit organizations, and NGOs. Foundations are among the world's leading institutional investors. The incorporation of environmental, social, and governance (ESG) factors into investment management has been described in a variety of ways: social investing, ethical investing, socially responsible investing (SRI), values-based investing, mission-related investing, sustainable investing, double- or triple-bottom-line investing, and responsible investing.

> With nearly $500 billion in assets in the United States alone at the start of 2009, foundations now hold a considerable stake in corporate America and the global economy.
>
> Like other institutional investors, such as pension funds, university endowments and religious groups, most foundations tend to be invested for the long term. Foundations are distinguished from many other institutional investors, however, by their explicit philanthropic missions. Since their emergence in the 19th century, foundations have dedicated their resources to tackling some of the most difficult social and environmental problems in communities throughout the United States and around the world.
>
> In many foundations, however, the values that drive grantmaking programs are still commonly separated from the financial management of foundation assets.
>
> At the same time, there has been a growing recognition that many mission-related social, corporate governance and environmental issues can be sources of financial risks and opportunities for foundation asset management.
>
> Mission-related responsible investing can provide foundations with several dynamic strategies that leverage their assets more fully for their core philanthropic purpose while managing risk and creating lasting value in their investment portfolios:

• **SOCIAL AND ENVIRONMENTAL SCREENING:** the practice of evaluating investments using social and/or environmental criteria in addition to traditional financial analysis.

• **SHAREOWNER ADVOCACY:** the actions taken by many socially aware investors in their role as company owners to dialogue with companies on social, environmental and/or corporate governance issues, and to file or co-file and vote on shareholder resolutions.

• **COMMUNITY INVESTING:** investments that direct capital to communities that are underserved by traditional financial services. Community investments provide access to credit, equity, capital, and basic banking products to low-income and marginalized regions in the US and around the world.

• **SOCIAL VENTURE CAPITAL:** typically debt or equity investments in early-stage for-profit companies that produce social and environmental benefits or support to nonprofit social enterprises.

. . . .

Regardless of the particular emphasis, in today's era of more "engaged" philanthropy, with "venture philanthropists" seeking more entrepreneurial, market-based solutions to social and environmental problems, social investing strategies have become increasingly embraced by many foundations seeking to leverage the full range of assets at their disposal.[1]

These corporate citizenship initiatives have resulted in a responsive change in multinational corporations. Many have adopted sustainability as a business philosophy and long-term strategy to create shareholder value. Companies such as General Electric, with its Ecomagination initiative, embrace opportunities and manage risks deriving from economic, environmental, and social developments. They focus on the potential for sustainable products and services, reducing or avoiding costs and risks that come with previous social and environmental practices.

The companies that are competent in meeting global and industrial challenges make good long-term investments. They remain competitive and enhance the reputation of their brands by integrating sustainability into their business strategy. They meet shareholder expectations for appropriate returns on investment, growth, and transparency. They use resources—financial, social, and environmental—to reinforce customer relationships through innovative products and services. They adopt ethical practices and invest in the development of their workforce.

Corporate executives no longer see, as Milton Friedman did in 1970, their moral mandate to "make as much money for their stockholders as they can within the limits of the law and ethical custom."[2]

"Put simply, corporate citizenship is the acceptance of the corporation's role as a responsible and significant member of its community."[3]

Guideline F Notes

1. *The Mission in the Marketplace: How Responsible Investing Can Strengthen the Fiduciary Oversight of Foundation Endowments and Enhance Philanthropic Missions* (Social Investment Forum Foundation, Washington, DC, 2009), www.socialinvest.org.
2. Milton Friedman, "The Social Responsibility of Business Is to Increase Its Profits," *The New York Times Magazine*, September 13, 1970, 32–33, 122–26.
3. Michael B. Goodman, *Corporate Communication for Executives* (Albany, NY: SUNY Press, 1998), 113.

Table F.1 Corporate Citizenship Organizations

NAME OF ORGANIZATION	MISSION	ADDRESS	INTERNET (URL)
Business Enterprise Trust (Nonprofit)	Promotes social leadership in business by recognizing honesty, integrity, and vision in the workplace. Award program. Case studies. Videos.	Business Enterprise Trust 706 Cowper St. Palo Alto, CA 94301 U.S. Ph. 650.321.5100 Fax 650.321.5774 Email bet@betrust.org	www.betrust.org
Business for Social Responsibility & BSR Education Fund (Nonprofit)	U.S.-based global resource for companies seeking to sustain their commercial success in ways that respect people, communities, and the environment.	Business for Social Responsibility 609 Mission St. San Francisco, CA 94111 U.S. Ph 415.537.0890 Fax 415.537.0888 Email membership@bsr.org	www.bsr.org
Business in the Community (Nonprofit)	Inspire business to increase the quality and extent of their contribution to social and economic regeneration by making corporate social responsibility an essential part of business excellence.	Business in the Community 44 Baker St. London W1M 1DH U.K. Ph. 44.0171.224.1600 Fax 44.0171.486.1700	www.bitc.org.uk
The Conference Board	Improve the business enterprise systems and enhance the contribution of business to society.	The Conference Board 845 Third Ave. New York, NY 10022 U.S. Ph 212.759.0900 Fax 212.980.7014 Email info@conference-board.org	www.conference-board.org
Council on Economic Priorities	Rates companies on social performance on topics such as environmental stewardship and treatment of employees; serves as a clearinghouse on issues of corporate responsibility.	Council on Economic Priorities 30 Irving Place, 9th Floor New York, NY 10003 Ph 212.420.1133 Fax 212.420.0988 Email cep@echonyc.com	www.cepnyc.org
Instituto Ethos (Ethos Institute of Business and Social Responsibility)	Promote the development of corporate social responsibility in Brazil.	Instituto Ethos Rua Francisco Leitao 469—conj. 1407 CEP 05414-020 São Paulo, Brazil Ph/Fax (55 11) 3068 8539 Email ethos@ethos.org.br	www.ethos.org.br
European Business Network for Social Cohesion (EBNSC)	To encourage and help companies to prosper in ways that stimulate job growth, increase employability, and prevent social exclusion, thereby contributing to a sustainable economy and a more just society.	European Business Network for Social Cohesion (EBNSC) Rue du Prince Royal 25 B-1050 Brussels Belgium Ph 32.2.502.83.54 Fax 32.2.502.82.58 Email ebnsc@ebnsc.org	www.business-cohesion.org

NAME OF ORGANIZATION	MISSION	ADDRESS	INTERNET (URL)
Interfaith Center on Corporate Responsibility (Coalition of 275 Catholic, Jewish, and Protestant institutional investors)	To hold corporations accountable on a wide range of social issues; shareholder resolutions, corporate meetings, letter-writing campaigns, and consumer boycotts.	Interfaith Center on Corporate Responsibility 475 Riverside Dr., Room 550 New York, NY 10115 U.S. Ph 212.870.2936 Fax 212.870.2023 Email info@iccr.org	www.iccr.org
Investor Responsibility Research Center	Impartial research for institutional investors on social issues, the environment, and corporate governance.	Investor Responsibility Research Center 1350 Connecticut Ave. NW, Suite 700 Washington, DC 20036 U.S. Ph 202.833.0700 Fax 202.833.3555	www.irrc.org
Keidanren	For a resolution of the major problems facing the business community in Japan and abroad, and to contribute to sound development of the Japanese and world economies.	Keidanren Kaikan 1-9-4, Otemachi Chiyoda-ku, Tokyo 100-8188 Japan Ph 81.3.3279.1411 Fax 81.3.5255.6255	www.keidanren.or.jp
Prince of Wales Business Leaders Forum	Promote socially responsible business practices that benefit business and society and which help to achieve social, economic, and environmentally sustainable development in emerging and transition economies.	Prince of Wales Business Leaders Forum 15-16 Cornwall Terrace Regent's Park London NW1 4QP U.K. Ph 44.0.171.467.3656 Fax 44.0.171.467.3610 Email info@pwblf.org.uk	www.pwblf.org
Social Investment Forum	To promote socially responsible investing.	Social Investment Forum 1612 K Street, NW Ste 650 Washington, DC 20006 U.S. Ph 202.872.5319 Fax 202.822.8471	www.socialinvest.org
Social Venture Network	A membership organization of successful business and social entrepreneurs dedicated to changing the way the world does business so as to create a more just, humane, and sustainable society.	Social Venture Network P.O. Box 29221 San Francisco, CA 94129 U.S. Ph 415.561.6501 Fax 415.561.6435 Email svn@wenet.net	www.svn.org

Guideline G: Core Competencies for Corporate Communication

Writing remains the core skill for corporate communication. Essential, too, is a thorough knowledge of the company and of business principles. The core competencies focus on:

- **Technical skills**: demonstrate strength in written and oral communication, as well as the ability to obtain information, critique, edit, and communicate effectively
- **Product/Market Knowledge**: shows an understanding of the customer, product, and market to effectively develop and deliver communications
- **Business Leadership**: demonstrates knowledge of business complexities and shows an ability to act as a strategic partner to leverage power of communication to drive business
- **The Industry Sector Environment**: demonstrates knowledge of the business and related legal, regulatory, and policy requirements
- **Reputation Management**: demonstrates the ability to build public trust and enhance the reputation of the business

For success as a corporate communicator in a global business environment, the skill set necessary also includes:

• integrity and honesty	• resilience
• global mindset	• positive attitude
• objective perspective	• energy, discipline, passion
• business orientation	• leadership
• project management	• team player
• critical and analytical thinking	• intelligent
• problem-solving, synthesizing	• innovative
• communication and media skills	• creative
• listening	• social ability
• persuasion	• emotional intelligence
• "grace under pressure"— confidence, composure, compassion	• mentoring and coaching
	• quick study
• cultural sensitivity	• strategic thinking

And these capabilities are needed for life-long learning in a global environment:

- Communication in the mother tongue
- Communication in foreign languages
- Mathematical competence and basic competencies in science and technology
- Digital competence
- Learning to learn
- Social and civic competencies

- Sense of initiative and entrepreneurship
- Cultural awareness and expression

Guideline H: Investor Relations

According to the National Investor Relations Institute:

> Investor relations is defined as a strategic management responsibility that integrates finance, communication, marketing and securities law compliance to enable the most effective two-way communication between a company and the financial community and other constituencies, which ultimately contributes to a company's securities achieving fair valuation.[1]

Sample Investor Relations Communication Plan

The creation of an Investor Relations Communication Plan implies a development process that includes: (1) setting long- and short-term goals and objectives; (2) determining IR responsibilities; (3) identifying and nurturing internal and external stakeholder relationships; (4) identifying key strategic and tactical elements of the IR program—calendar of required events and filings; (5) establishing an annual budget; (6) determining quantitative and qualitative measures for success. The plan itself includes the six sections outlined below.

1. Executive Summary
2. Goals:
   Attract longer-term investors
   Stabilize investor base
   Build value for shareholders
3. Target highest potential institutional investors for one-on-one marketing:
   Concentrate on small- to mid-cap, value-oriented portfolios
   Highly selective approach
   Requires management accessibility
4. Market aggressively to individual investors:
   Market through ____
   Capitalize on consumer appeal & familiarity
5. Increase analyst coverage:
   Target brokerage houses with strong institutional sales
   Address industry analysts to build credibility
6. Program:
   Schedule: ___ months
   Budget: $____
   Staff & Reporting Structure

Sample Investor Fact Sheet (one double-sided sheet)

Description of the Business
· Global, comprehensive, and broadly based ____ of ____
· Serves the ____ markets with a focus on ____ products

Consistent Performance
· ____ years of sales increases
· ____ years of earnings increases adjusted for special charges
· ____ consecutive years of dividend increases
· Annual compounded growth achievements:

	One Year	Five Years	Ten Years
Sales			
Earnings per Share (GAAP)			
Dividend Growth			
Total Return to Shareholders			

Broadly Based/Exceptional Financial Strength
· Over ____ operating companies in ____ countries selling products throughout the world (__% of sales outside of United States)
· Sales split among ____ business segments
· More than ____ marketed; ____ over $__ million; __ over $____ million; __ over $__ billion
· One of a few U.S. industrial companies that still commands a Triple A credit rating by both Standard & Poor's and Moody's credit rating agencies
· Generated approximately $____ billion annual free cash flow in 20__

Worldwide Market Leadership
· Largest ____ company
· __th largest ____ company globally
· Leadership positions in major markets: (list of markets)
· Approximately ____% of sales derived from products/businesses that have a #1 or #2 global market share position

Relevant Charts such as:
· Net Earnings over 10 years
· Shareholder Return Comparison over 5 years with S&P Index, Industry Specific Indexes
· 20____ Sales by Geographic Area, for example, US, EU, Asia Pacific
· 20____ Sales Segment, for example, consumer, manufacturing

· New Trade Sales over 10 years

side two

Strong Commitment to Research and Development
· $____ billion in Research and Development spending in 20____
· Approximately ____+ external alliances and collaborations entered into annually
· Approximately ____% of products sold in 20____ introduced in last 5 years

Positioned for Future Growth
· Solid pipeline of new and innovative products
· U.S. demographics point to high growth in ____ industry sector
· Broad geographic reach provides ability to rapidly introduce new products in markets
  around the world
· Focus on products that reduce cost
· Management based on a core set of Strategic Principles

.Investor Information

NYSE Symbol: _____

For additional information, contact:
Investor Relations Department
Address
City, State, Zip
Voice: (800) ____-_____
Fax: (____) ____-_____

Press releases, including earnings reports, are available via web access
COMPANY WEBSITE

For company reports, including Annual Reports and SEC filings, visit our homepage at:
  COMPANY WEBSITE

The Annual Report

The annual report is the company report card that provides information and data about the company's accomplishments, progress, business mission, and management philosophy for government regulators, the investment community, shareholders, employees, and the general public. An indirect but essential goal of the annual report is its role in the creation and perpetuation of the identity, reputation, and image of the organization. Because it is often the first document consulted when researching a company, it is also used to persuade individual and institutional investors that the company is a sound place in which to put capital. It has a long shelf life, which makes it one of the most important documents a company produces.

Copies of the report go not only to all registered stockholders, but also to Wall Street analysts, the business media, libraries, vendors, trade associations, students, and professional groups. It is often a requirement in new business proposals to clients and the government, and it is frequently used for recruiting new employees.

For these reasons, and because the SEC requires publication of much of the information in the annual report, companies take the development of an annual report seriously. Every element of the annual report is designed to contribute to the positive image of the company: the artful covers, professional photographs, drawings, charts, graphics, the paper it is printed on, as well as the web and digital media versions of the annual report.

Components of the annual report:

The Auditors' Report is a summary of the findings of an independent firm of certified public accountants showing whether the financial statements are complete, reasonable, and prepared in a way consistent with generally accepted accounting principles (GAAP) at a set time.

Management Discussion & Analysis (MD&A) is a series of short, detailed analytical reports on thcompany's performance, covering operations and the adequacy of liquid and capital resources to fund those operations.

Financial statements and notes are statements that provide the raw numbers for the company's financial performance and recent financial history. The SEC requires statements of:

- earnings
- financial position
- cash flows

These statements also include comprehensive notes, explanations, additional detail, and supplementary financial information.

Selected financial data offers information that summarizes a company's financial condition and performance over five years or longer. Data for making comparisons over time may include revenue (sales), gross profit, net earnings (net income), earnings per share, dividends per share, financial ratios such as return on equity, number of shares outstanding, and the market price per share.

Compensation Discussion & Analysis (CD&A) In 2006 the U.S. Securities and Exchange Commission required a section titled Compensation Discussion and Analysis, addressing the company's policy and rationale for executive pay. The CD&A covers topics concerning the company's top five executive officers' pay—generally the CEO, the CFO, and the three other most highly paid executives. The discussion covers the compensation program's objectives, what the program is designed to reward, the specific elements of compensation, why the company chose those items, how much the company allocates for each item, and the decision-making process for overall compensation objectives.

Financial highlights is one of the most-read sections of an annual report. It contains highlights that give a quick summary of a company's performance. The numbers appear in a short table and supporting graphs.

Letter to stockholders may be from the Board of Directors, the CEO, or both. It can provide an analysis and a summary of the year's events, including any problems, issues, and successes the company had. It usually reflects the business philosophy and management style of the company's executives and often lays out the company's direction for the next year.

Corporate message may be seen by some analysts, business executives, and stockholders as an advertisement for the company. It portrays the company in the way it would like to be perceived by others through the photographs, illustrations, and narrative. The discussion can cover the company's business units, markets, mission, management philosophy, corporate culture, and strategic direction.

Report of management usually is a letter from the Board Chairperson and the CFO, takes responsibility for the validity of the financial information in the annual report and states that the report complies with SEC and other legal requirements. The discussion attests to the presence of internal accounting control systems that cover effectiveness of operations, reliability of financial reporting, and compliance with federal laws.

Board of directors and management lists the names and titles of the company's board of directors and top management team. Sometimes companies include individual and group photographs.

Stockholder information covers the basics: the company's corporate office headquarters, the exchanges on which the company trades its stock, the location and time of the next annual stockholders' meeting, and other general stockholder service information.

Guideline H Note

1. *NIRI Standards of Practice for Investor Relations*, 3rd ed., January 2004, 5; www.niri.org.

Guideline I: Transparency and Disclosure

Material Information

According to securities laws, information is *material* if its disclosure is likely to have an impact on the price of a security, or if reasonable investors would want to know the information before making an investment decision. Positive and negative information can be material, as well as a forecast of an event that may or may not occur. If there is a doubt about the materiality of information, most IR professionals suggest disclosure.

Examples of material information about a company can include: a major change in senior management; the launch of a new product or business; announcements of earnings or losses; a change in earnings or in forecast earnings that is higher or lower than the forecast; a pending merger or acquisition; sale or purchase of significance; gain or loss of a substantial customer or supplier.

Plain English and Disclosure

Warren Buffett wrote in his preface to *A Plain English Handbook* (1998):

> For more than forty years, I've studied the documents that public companies file. Too often, I've been unable to decipher just what is being said or, worse yet, had to conclude that nothing was being said. If corporate lawyers and their clients follow the advice in this handbook, my life is going to become much easier.

> There are several possible explanations as to why I and others sometimes stumble over an accounting note or indenture description. Maybe we simply don't have the technical knowledge to grasp what the writer wishes to convey. Or perhaps the writer doesn't understand what he or she is talking about. In some cases, more-

over, I suspect that a less-than-scrupulous issuer doesn't want us to understand a subject it feels legally oblig-
ated to touch upon.

Perhaps the most common problem, however, is that a well-intentioned and informed writer simply fails
to get the message across to an intelligent, interested reader. In that case, stilted jargon and complex con-
structions are usually the villains.

This handbook tells you how to free yourself of those impediments to effective communication. Write as
this handbook instructs you and you will be amazed at how much smarter your readers will think you have
become.

One unoriginal but useful tip: Write with a specific person in mind.

When writing Berkshire Hathaway's annual report, I pretend that I'm talking to my sisters. I have no trouble picturing them: Though highly intelligent, they are not experts in accounting or finance. They will understand plain English, but jargon may puzzle them. My goal is simply to give them the information I would
wish them to supply me if our positions were reversed. To succeed, I don't need to be Shakespeare; I must,
though, have a sincere desire to inform.

No siblings to write to? Borrow mine: Just begin with "Dear Doris and Bertie."[1]

Interactive Information and XBRL

From the January 2009 *Toward Greater Transparency: Modernizing the Securities and Exchange
Commission's Disclosure System—21st Century Disclosure Initiative: Staff Report*:

> The mission of the Securities and Exchange Commission is to protect investors, maintain fair, orderly, and
> efficient markets, and facilitate capital formation. An important tool supporting this mission is disclosure.
> To fulfill its function, the disclosure system provides for full, fair, and timely disclosure of material infor-
> mation that is accessible and easy to use. . . .
>
> . . . too much legal or technical jargon gets in the way of clarity, the documents can be too long and wordy,
> and information is sometimes hard to find in the reports. Not surprisingly, [investors] turn to financial advi-
> sors or brokers as their most important source of investment information. So it is not that the disclosure
> documents contain unimportant information; it is that the time required to extract the relevant informa-
> tion makes such activity too burdensome for many other than the professional investor. This Initiative seeks
> to find ways to help investors access relevant disclosure materials with less effort and cost, while also mak-
> ing them easier to use. In this sense, we seek to democratize investment information.[2]

Plain language is a powerful and essential element in transparency and disclosure. And the SEC's
2009 requirement to use XBRL is another:

> We are adopting rules requiring companies to provide financial statement information in a form that is
> intended to improve its usefulness to investors. In this format, financial statement information could be
> downloaded directly into spreadsheets, analyzed in a variety of ways using commercial off-the-shelf soft-
> ware, and used within investment models in other software formats. The rules will apply to public compa-
> nies and foreign private issuers that prepare their financial statements in accordance with U.S. generally
> accepted accounting principles (U.S. GAAP), and foreign private issuers that prepare their financial state-
> ments using International Financial Reporting Standards (IFRS) as issued by the International Accounting
> Standards Board (IASB). Companies will provide their financial statements to the commission and on their
> corporate websites in interactive data format using the eXtensible Business Reporting Language (XBRL).
> The interactive data will be provided as an exhibit to periodic and current reports and registration state-
> ments, as well as to transition reports for a change in fiscal year. The new rules are intended not only to make

financial information easier for investors to analyze, but also to assist in automating regulatory filings and business information processing. Interactive data has the potential to increase the speed, accuracy, and usability of financial disclosure, and eventually reduce costs.[3]

Sample Disclosure Policy

Statement of Purpose presents the company disclosure policy and its contacts with the investment community.

Designation of Authorized Spokespersons lists the company's authorized spokespersons, usually the CEO, CFO, and IRO (Investor Relations Officer), who know the company's financials, understand disclosure rules, and can respond to questions about the company's stock and performance.

Duties of Employees are to refer any calls from financial analysts and business reporters to the authorized spokespersons. They are instructed not to discuss non-public information about the company with anyone. Also, many company CFOs remind all employees of this policy one month before the company earnings release date.

Speakers from the Financial Community are invited through Investor Relations to present at company meetings, except in the weeks before earnings are released.

Making Earnings Projections is prohibited. General guidance about expected sales growth and operating expenses is only offered during the quarterly earnings conference call. No forward-looking financial information that conflicts with the company's SEC filings is given.

Responding to Analysts' Projections: Generally the policy is to not specifically endorse external estimates, and under no condition to comment on a particular earnings estimate.

Reviewing Analysts' Reports to correct only factual information that is inaccurate. Do not endorse analyst conclusions, particularly earnings forecasts, financial projections, or recommendations. Do not confirm or deny any of the reports' statements regarding predictions or projections or confirm the accuracy of the earnings models.

Commenting on Rumors instructs everyone in the company about the policy of not responding to rumors originating outside the company.

Conference Calls and Webcasts are held after the release is issued and after the close of the market in conjunction with quarterly earnings releases. They are broadcast live by phone as well as webcast from the company Internet site. The call is open to all investors and is available for 48 hours after the call. The webcast is available on the website for the quarter.

Investor Conferences, Analyst and Shareholder Meetings are webcast live on the Investor Relations website.

Guidance: Detailed P&L guidance, including a sequential revenue growth range as well as operating expense guidelines, tax rate, and share-count expectations, are provided during quarterly earnings release conference calls. Additionally, a business update is issued in the form of a press release during the third month of each quarter, usually after the market close. The exact date and location of these updates is published in the quarterly earnings release. Additionally, these updates are available via "push" technology by signing up for a financial press release alert on the IR web site. The business update consists of several bulleted statements that are qualitative in nature, and none are believed to be material by the company. There are certain items that are not discussed with the investment community, including bookings numbers, backlog figures, design wins per quarter, actual pricing, gross margins, by-product line, specific product, or customer revenue.

Quiet Period is the time following the business update during which the company does not discuss with investors new information related to the current quarter's performance until the issuance of the quarter-end earnings release. It is intended to avoid the potential of selective disclosure and the spread of rumors before the earnings announcement.

Selective Disclosure is to be avoided, since the IR goal is to treat all investors fairly and equally. The policies for the conference call and guidance are intended to prevent selective disclosure.

World Wide Web is treated as a site for publicly disseminated information about the company.

Online Chat Rooms can be monitored for what others are saying about the company on the World Wide Web. Most companies prohibit employees from discussing business information that belongs to the company on Internet chat rooms during working hours and using company computer systems. Most companies do not respond to rumors or correct any inaccuracies that might appear in online chat rooms. Employee participation in chat rooms may compromise sensitive company information and violates most company disclosure policies. Misuse of Internet chat rooms may result in termination of employment.

Insider Trading Policy is maintained and distributed by the company legal department.

Sample IR Communication Policy Distributed Quarterly

Investor Relations and Public Relations teams manage the company's communications with the financial community, the press, and research analysts.

If you are contacted by an investor or any member of the financial community, please immediately refer them to a member of the Investor Relations team. Do not answer any questions or provide any information regarding the company or its business.

Employees who violate this policy may be subject to disciplinary action, including immediate termination, as well as possible prosecution for violation of securities laws.

If you are contacted by a member of the press or a market analyst, please contact the Public Relations team immediately. This policy also applies if you are asked to participate in or coordinate a keynote speech or panel being held in at an industry event or trade show.

Dos and Don'ts When Speaking to Financial Analysts and Investors

Don't say anything that you don't want to see in print.
Don't say anything that you don't want your competitors to know.
Don't talk about earning projections.
Don't talk about how the quarter is looking.
Don't discuss any non-public material information.
Don't talk about revenues coming from any specific customer.
Don't meet with any financial analysts without an investor relations person present.
Don't comment or respond to any Internet chat rooms.
Always take the opportunity to put the company in a favorable light with facts, not hype.
Use market research forecasts—not internal projections—for charts.
Always state the source when using any forecast.
Inform Investor Relations if you receive a call or request from the financial community.
Schedule any meetings with financial analysts through investor relations.

Guideline I Notes

1. Warren Buffett, *A Plain English Handbook: How to Create Clear SEC Disclosure Documents* (Office of Investor Education and Assistance, U.S. Securities and Exchange Commission, Washington, DC).
2. *Toward Greater Transparency: Modernizing the Securities and Exchange Commission's Disclosure System—21st Century Disclosure Initiative: Staff Report* (Washington, DC: US SEC, 2009), 3.
3. U.S. Securities and Exchange Commission, "Interactive Data to Improve Financial Reporting," 17 CFR Parts 229, 230, 232, 239, 240, and 249.

Guideline J: Reputation Management

Over the past two decades, most multinational corporations have established a formal or semi-formal system of reputation management. However, there is a wide divergence in how individual companies define reputation management and the processes they use for doing so. Although it is not the most commonly observed model, we believe that "Reputation Management" properly includes both defensive and offensive elements. By defensive, we mean the methods (some of which we described under "Crisis Communication," above) by which companies avoid negative reputation impacts by carefully monitoring for and preparing to deal with emerging issues. By offensive, we mean all those methods that companies use to enhance their corporate reputations.

Since we deal with issues management specifically in Guideline B, we will only briefly summarize the elements that go into defensive reputation management. These include:

1. Proactive monitoring of issues such as the environment, human rights, or corporate governance in which changing trends could cause the company's policies to be singled out for criticism
2. Building and maintaining relationships with NGOs and other potential critics so that the company's positions are well understood, and that it gets early warning of changing trends
3. Designing and implementing an effective issues response system to ensure that crises are quickly and expertly managed

The quality of the offensive component of a company's reputation management effort is usually a direct result of how rigorously those efforts are measured and how specific they are in seeking to achieve meaningful goals. Reputation management targets should be designed for each key stakeholder audience:

- Employees
- Customers/consumers
- Shareholders
- Communities
- Business partners
- Regulators

While reputation in the aggregate might appear to be an intangible quality, the more specific the desired outcomes with respect to an individual stakeholder group, the more powerful the potential effect. For example, a strong reputation for product innovation with business partners is a generally positive outcome, but much more powerful is seeking to increase the likelihood that a business would look positively on a joint venture because of the original company's talent for innovation. Similarly, most companies attempt to improve their standings in surveys that purport to measure how positively employees evaluate them as a "good place to work." More specific metrics, however, such as the perceived value of the company's training, its working environment, and its career-building culture are more likely to support positive business outcomes such as strength of recruitment and low turnover.

Here are some of the criteria used to measure corporate reputation:

- Easy to do business with
- Consumer-friendly
- Fiscally well-managed
- Long-term growth oriented
- Product/Service innovation
- Ethics
- Socially responsible
- Long-term investment
- Globally competitive
- People management
- Quality of products/services

Reputation Management Campaigns

It is very important that reputation management not become a passive box-checking exercise. While it is clearly of some value to management leaders to be able to show their boards of trustees improving numbers on widely read reputation lists such as *Fortune*'s "Most Admired Companies," we believe performance improvement in reputation requires a more hands-on approach.

That is why we recommend a "campaign" approach to reputation. What the campaign approach (which should be renewed annually) achieves is that it forces those people responsible for reputation in the corporation to be very specific about their objectives. Putting a calendar time-frame in place for the campaign helps make this possible. Reputation management can be put in place for the coming fiscal year as part of the same corporate planning cycles for revenue growth, new product launches, and profitability targets. Having annually revised reputation goals also helps to keep such programs fresh and attuned to changing market realities.

Reputation Measurement Systems

Various reputation measurement metrics are in use today. Most of them are based on asking a select subgroup of stakeholders their opinion about a range of the company's activities. Responses, both positive and negative, are then scored on a fixed scale, resulting in a weighted number for an aggregate reputation score. Quantitative opinion metrics are frequently augmented with focus group research. Other systems use market value (i.e., shareholder opinion) as a proxy for the opinion of all stakeholders.

As Table J.1 shows, there is a wide variety of reputation methodologies, and a clear understanding of the assumptions underlying the methodology selected is crucial. Whichever method is deployed, it should be actively connected with the annual campaign designed to improve specific reputation criteria.

TABLE J.1 Reputation Measurement Systems

Source	Name	Respondents	Criteria
Financial Times	World's Most Respected Companies	4,000 business leaders in 70 countries	Shareholder value, corporate social responsibility, integrity, corporate governance
Millennium Poll	Prince of Wales Business Leader Forum and The Conference Board	25,000 average citizens in 23 countries	Corporate Social Responsibility
Fortune	America's Most Admired Companies	Poll of 10,000 executives, directors, and analysts	Innovation, quality of management, people management, use of corporate assets, social responsibility, global competitiveness, financial soundness, long-term investment, product/service quality
Harris Interactive	Reputation Quotient*	Poll of 22,000 members of the general public	Financial performance, vision & leadership, products & services, workplace environment, emotional appeal, social responsibility
Y&R	Brand Asset Valuator	230,000 consumers in 44 countries	Differentiation, Relevance, Esteem & Knowledge
Reputation Institute	Global Reputation Pulse	60,000 online interviews with the general public in 25 countries	Leadership, citizenship, governance, workplace, products/services, innovation, performance
Landor	Image Power— survey of top 250 technology brands	11,000 Internet users in 10 countries	Corporate reputation, financial performance, name-brand recognition, company expertise, advertising, online presence, CEO reputation/personality
CoreBrand	Brand Power and Brand Equity rankings	Survey of 12,000 senior executives at large companies	Familiarity + favorability as measured by overall reputation, perception of management, investment potential

A number of other business media, such as The Wall Street Journal, publish their own reputation surveys based on the Harris Interactive Reputation Quotient model.

Guideline K: Transaction Communication

Transaction Communication Definition: The phrase "Transaction Communication" refers to written, oral, electronic, or any other communications relating to the proposed merger transaction that are transmitted to a large audience, whether internal or external (including, without limitation, video replays). It is not intended to cover daily communication among co-workers.

During times of uncertainty, which are part of company mergers and acquisitions, it is especially important to communicate with employees to provide current information and help everyone stay focused on his or her role in achieving objectives. It also can be important to communicate externally with key business partners, opinion leaders, and consumers. Care must be taken in these communications to follow leading practices, as well as to account for some special considerations that apply in the context of Transaction Communications.

The following guidelines were developed by the Corporate Affairs staff at a publicly traded company, with assistance from the Office of Integration Planning and the Human Resources and Law Departments, to assist leaders and communication professionals in their communication efforts.

Key Considerations in Preparation of Transaction Communication

In preparation of any Transaction Communication, you should be mindful that:

- It is essential that leaders and communication professionals serve as a source of credible, accurate information for employees and others during a period of uncertainty.
- [Company Y] and [Company X] will continue to be competitors prior to the closing of the merger, and communications should not imply otherwise.
- Absent specific authorization from the Office of Integration Planning or from a Management Committee member, [Company Y] employees may not communicate with [Company X] employees, including communication professionals.
- Special legal considerations apply to Transaction Communication. Considerations are complex and time sensitive and may require advance review of certain materials by the Law Department.

General Guidelines

- Follow responsible business communication practices.
- Acknowledge what is not known, and avoid engaging in speculation.
- Direct individuals to the HR FAQs on the Transition Update website and their HR business partner for answers to questions regarding employment, compensation, and benefits.
- Use the pre-approved communication materials in the Communication Library area of the Transition Update website for the specific audience (internal or external) for which they were approved.
- When using materials from the Communication Library that includes legends, ensure that you continue to use the legends.
- Note that certain materials in the Communication Library are available by request only, as Corporate Affairs must keep track of who receives these materials and how/when they are used.
- Do not create new websites or sub-websites that include content relating to the transaction; rather, employees should be directed to the Transition Update website. The only exception is that existing materials on the Transition Update website may be translated into other languages and posted on a local website as needed.
- The Transition Update website will be updated to include information on integration teams

and activities. This will be the appropriate vehicle for new information from the teams as it becomes available.

- It is preferable to avoid creating additional original content for communications about the transaction. If you need new content or need to significantly change an existing communication, contact [name] so that proper review and approval can occur before use. All new material must initially be prepared in English.
- Do not share communications and messages marked for "Internal Use Only" outside the company.
- All media contact should be referred directly to [Media Relations]. They will coordinate any necessary response.

Review Process for New Content in Transaction Communications

Corporate Affairs, working with the Office of Integration Planning and the Law Department, must review all Transaction Communication that includes significant new content relating to the Transaction (i.e., content not included in the pre-approved materials in the Communications Library on the Transition Update website) *before* it is distributed to the intended audiences.

- Corporate Affairs will review Transaction Communications for accuracy and then submit them, as appropriate, to the Office of Integration Planning, which will review the communications if they relate to integration activities, and the Law Department, which will review the communications for legal compliance, including whether any legends or public filings are required under the securities laws.
- Transaction Communications that are approved may be translated into other languages at a local level.
- Use the following approval process for Transactions Communication with significant new content. (Again, this is not necessary if you use the materials in the Communications Library on the Transition Update website.)
  —*Step 1:* Submit all Transaction Communication with significant new content in draft form by email to [Name] of Corporate Affairs for review and routing to the Office of Integration Planning and the Law Department as appropriate. Please anticipate at least 48 hours for Corporate Affairs, Office of Integration Planning, and Law Department review. We recognize that there will be urgent, unanticipated requests for review, and we will do our best to accommodate them. In the event that your Transaction Communication is urgent, please call [Name] to alert her to the situation, in addition to sending your request to her by email. Try to limit urgent requests.
  —*Step 2:* [Name] will return reviewed communications to the originator, who is responsible for making revisions requested during the review process.
  —*Step 3:* The originator must submit the final Transaction Communication to [Name] before it is first used (and in any event by no later than 12:30 P.M. on the date of first use), with a copy to Legal. This is necessary to ensure that any securities-law filing requirements can be met. It will also enable Corporate Affairs, where appropriate, to add this now-approved content to the Communication Library on the Transition Update website for future use by others in the company.

Contacts: Your Questions and Media Inquiries

If you have any questions regarding your role in complying with [Company Y]'s approval process or legal requirements, please contact the following individuals: [Add names].

Guideline L: Affiliate Relations

In its Pilot Study of Affiliate Relations, CCI Corporate Communication International asked 12 selected multinational corporations to respond to a six-question online survey. Seven corporations answered the questions. Two corporations in our sample had a specific management function dedicated to Affiliate Relations, or a similar role. One commented, "We do have a colleague who is responsible for international communications, and is first point contact for our affiliates. They come to her for any advice they need about corporate and product communications in particular."

Five of the respondents currently consider it a priority to communicate with local affiliates. Five consider the need for affiliate relations to be increasing, one decreasing, and one staying the same. One commented, "International business [is] growing faster." Another commented, "Currently, our company only does business in two countries outside the US. We anticipate that number will increase in the future." One underscored the growing importance by saying that the "need for consistency in messaging and globalization of the communications function becomes clearer. This topic is also higher on the agenda of the Board. Importance is understood well nowadays."

Four of our respondents did address communication with local affiliates, but not through a dedicated position. One corporation added, "We do this for communication about specific functions and internal communications. In those areas, specialists in specific fields will reach out to key stakeholders within local affiliates." And another said that communication was "through regular meetings and message distribution."

Three of our respondents used an extensive feedback loop with local affiliates to ensure quick answers to questions, while one said that not enough was being done to ensure consistency in company messaging. One added, "We require alignment with corporate messaging and review all broadband messaging in the businesses and regions. [We have] lots of centralized control."

Two used occasional audits of local and external messaging for affiliates to measure communication performance between headquarters and affiliates; two used periodic surveys; and two had no concrete ways in which to measure performance. One added, "Meetings (either face-to-face or video-conferencing) with local affiliates on a regular basis. Also weekly update meetings for all communication colleagues by phone." Another said, "regular/daily engagement."

Pilot Study Overview

To help corporate professionals understand the forces and trends influencing the practice of corporate communication, CCI has carried out studies of the current state of corporate communication in Fortune 1000 companies since 1999, as well as benchmark studies in China (2006 and 2008), South Africa (2008), the European Union (2009), and Australia and New Zealand (2009). Its next study of Fortune 500 Companies is set for 2009. As the corporate communication profession responds to a changing global business environment, internal communications has taken on a more important role. This is being called Affiliate Relations by selected multinational corporations.

This pilot study of the practice of Affiliate Relations polled corporate communication professionals in 12 publicly traded companies, using a six-question online survey through a secure website. The email message and the text of the six-question survey comprise Exhibit 1 below. The information-gathering phase was carried out over a one-week period ending December 24, 2008, a time in which many corporations take time off for the holidays. For this reason the period was extended to January 16, 2009. This report was made available to the companies participating in the pilot study. The results will be used as part of ongoing CCI projects to document corporate communication practices and trends. Responses to the six questions follow:

1. What does your company do to ensure that messages, materials, initiatives, and activities driven by headquarters are understood, utilized, and implemented in an aligned and effective manner by local affiliates? (Select all that apply.)

    50%   We have an extensive feedback loop with local affiliates to ensure that we are providing them with quick answers to any questions they might have.

    33%   We closely monitor messaging from each of our local affiliates.

    50%   We have periodic meeting sessions to reiterate core messaging.

    17%   We control messaging tightly from the top down.

    17%   We do not do enough to ensure consistency in our messaging.

            Other—Please specify

2. Do you have a specific function dedicated to such affiliate relations or a similar role?

    29%   Yes

    71%   No

3. Do you address communication with local affiliates, but not through a dedicated position?

    67%   Yes

    33%   No

4. How do you measure your performance in communication between headquarters and affiliates? (Select all that apply.)

    33%   Occasional audits of local internal and external messaging for affiliates.

    0%   Quarterly reports of messaging from local affiliates.

    33%   Periodic surveys on the degree of success of messaging consistency among headquarters and local affiliates.

    33%   We do not have concrete ways to measure our performance.

5. Do you currently consider it a priority to focus on addressing the need to communicate with local affiliates?

    83%   Yes

    17%   No

6. Do you anticipate the need for affiliate relations increasing, decreasing, or staying the same in the future?

    71%   Increasing

    14%   Decreasing

    14%   Staying the same

Guideline M: Thought Leadership and Executive Relationship Management

The role of Chief Executive Officer has evolved in many ways since the birth of the modern corporation. Starting with the role inherent in the title—execution of the company's strategy—the top job has also come to include being the face and voice of the organization for a range of stakeholders. In modern crisis management theory, the CEO's active presence is now understood to signal that the company is taking the matter seriously. In today's environment of rapid technological and economic change, the chief executive is also required to be the visionary for the company, explaining not just where the company is going but where the world is going. This role has come to be called thought leadership. At the same time, the primacy of the chief executive's position in a company, analogous to President Teddy Roosevelt's calling the White House a bully pulpit, means that every contact a chief executive makes is an opportunity to enhance the brand and strengthen relationships. In many companies, this opportunity is managed proactively and called "executive relationship management (ERM)." Thought leadership and ERM, taken together, represent the effective use of the chief executive's role to enhance the reputation of an organization with a wide range of stakeholders.

The technology revolution of the 1990s brought with it a critical shift in what companies (and, by extension, chief executives) needed to be knowledgeable about. Prior to this period, it was reasonable to expect that senior corporate officers would be deeply intimate with their own company and well informed about the history and evolution of the industry in which their company operated. When called upon to make speeches, the chief executive would speak conventionally about the challenges and achievements of his own organization, with perhaps a brief rhetorical flourish in the direction of broader issues. The impact of the information technology revolution of the 1990s was to give an extraordinary value to the ability to interpret events and to predict the next cycle of innovation. As the world became familiar with "enterprise resource planning" and "relational databases," as the Internet created and destroyed business models, it became increasingly important for companies to demonstrate to their stakeholders that they understood where the world was going and would not be left in a decaying backwater of their industry like the county seats that were bypassed by the railroads in the 19th century. At the same time, the public began making new demands on the corporation in the form of calls for greater engagement in what came to be known as corporate citizenship. Spurred by the consequences of rapid globalization, publics in the advanced economies wanted to know where corporate management stood on issues of diversity, human rights, fair trade, the environment, and energy use.

The interweaving of these two strands created an environment in which one of the key attributes of the successful CEO is the ability to offer the company's perspective on a variety of world issues. A well-managed thought leadership program is designed to deliver this perspective as effectively as possible.

Designing a Thought Leadership Program

While thought leadership is not the exclusive province of the chief executive, he or she must be able to feel ownership. This means that for a thought leadership program to succeed, it needs to be aligned to some degree with his or her personal interests and experience. This is one criterion for selecting the area of focus for a thought leadership program. Others include:

- Relevance to the core business or industry in which the company is engaged
- Visibility of the issue in the media and among think tanks
- The existence of a range of perspectives on the subject so that there is something legitimate to debate
- The perspective the company takes on the issue should not be exclusively self-serving

- The ability of the company not just to speak out on the issue but to convene a range of voices to produce new thinking
- The content should not be so technical as to exclude generalist understanding and input
- No other CEO in the same industry has selected the topic as a platform

Examples of thought leadership that fulfill these criteria are the support of innovation in evidence-based medicine by pharmaceutical companies; the debate about American higher education in the competitive global marketplace convened by technology companies; the issue of intellectual property rights to digital content on the Internet; privacy in e-commerce; the debate about the need for global financial regulatory authorities; how to create a viable marketplace for non-carbon-based fuels. Each of these issues has an element of controversy, is part of a current debate, and, to the extent that it is self-serving, is nonetheless of interest to a broader constituency.

In exploring potential issues on which to base a thought leadership program, the communication team should also:

- Research academic, governmental, and other thought leaders who could be engaged with the CEO on the topic in question
- Determine which business and academic venues are suitable for the discussion of the issue
- Identify reporters, bloggers, and columnists who have written about the issue
- Create a database of nonprofit organizations and other public entities for whom this issue is a key agenda item, whether or not they share the company's perspective
- Assess whether a corporate perspective on the issue could be too controversial or, conversely, too uninteresting

As the foregoing criteria indicate, the best thought leadership programs manage to strike a middle path between controversy that would be distracting and Motherhood and Apple Pie.

Executing the Thought Leadership Program

Once the issue has been selected, it is advisable to take some soft soundings with influential voices on the issue to ensure that the company's contribution to the debate would be welcome. Having passed this test, the program can begin in earnest. The communication channels to be deployed are limited only by the resources and the cultural comfort level of the company. The most commonly used tools in thought leadership are:

- Conference speeches
- White papers
- Op-ed pieces
- By-lined articles
- Advertorials
- Blogs
- Webcasts
- Podcasts/V-casts
- Web pages on the corporate site
- Convening conferences
- Research grants

- Association memberships
- Recognition awards and prizes for other advocates on the issue

Conventional measurement and tracking tools can then be deployed to measure the frequency with which the executive's name is picked up in association with the issue selected, and how the visibility of the issue as a whole is picked up in the traditional and web media. Where possible, a distinct verbal formulation of the issue or the executive's position will help identify how widely the issue has been picked up by others from this particular source. A well-executed thought leadership campaign contributes to the visibility of the organization and to its reputation as being committed to the resolution of important community issues.

Executive Relationship Management

An executive relationship management program can be carried out in the absence of a thought leadership platform, but the two concepts are immeasurably enriched when operated in tandem. In an executive relationship management program, the communications team develops a tool for creating ongoing opportunities for the CEO to interact with peers, customers, government officials, and other thought leaders. The list of relationships should begin with the CEO's existing contacts and be augmented with input from top executives and salespeople with insight into key accounts. The concept is to augment these relationships and increase the CEO's network by finding opportunities for personal outreach by the CEO. The nature of the contact should be determined by the CEO's personal preferences and the nature of the relationship but can include letters and emails, phone calls and cards.

Once the core list is in place, the communications team sets up news filters in a web-based monitoring service to capture in real time any news about the individual contacts in the database or issues which, based on the relationship, the team knows will affect that individual. The criteria for correspondence or contact can be personal milestones, key business or political wins, promotions, appointments, awards and recognition, important speeches, or significant coverage of the contact. It can also include important news coverage of a business or political event of mutual interest or concern.

It is important that the contacts be initiated in a timely fashion so that they are still fresh in relation to the events that prompt them. As the database and contact stream grows, the metrics to be tracked include:

- Growth in the size of the network
- Frequency of contacts
- Outcomes from contacts, such as responses, invitations, business opportunities, visibility
- Cross-fertilization of contacts that increase network density

The combination of thought leadership programming and executive relationship management is a powerful tool in nourishing corporate reputation and visibility using the role of chief executive as a valuable image conduit for the organization.

Guideline N: Social Media & Corporate Blogging

The use of social media by corporate communicators is still in its infancy as are many of the platforms themselves. Some of the ways in which these platforms are being used will stand the test of time, others will fade away and some uses, yet unknown, will surely emerge. Some guiding principles, have, however, begun to assert themselves. We will discuss blogging and working with bloggers in this guideline because we think that there are more similarities between bloggers and social media than between bloggers and traditional media. Others may disagree and there are certainly some aspects of relating to bloggers that mimic traditional media relations.

While there are dozens of social media networks, we believe the following platforms are the most important for US communicators—Facebook, LinkedIn, Twitter and YouTube. Although Facebook is a powerful presence around the world, there are many countries where it is dwarfed by other platforms, such as Orkut in Brazil, Mixi in Japan and 51.com and QQ in China. In South Korea, the country with the fastest broadband connections in the world, the leading social media platform is Cyworld.

Successful social media activity by corporations is based on some key principles:

1. Frequent and rich content creation and refreshment
2. The content is designed to take advantage of the specific features of the platform
3. Frequent, candid interaction with social media participants extended over time
4. The users, not the company, "own" the platform

Adhering to these principles requires a true commitment from senior management as this is not a small resource allocation. This means that, before embarking on a social media strategy, communicators need to create a plan for management that shows the benefits in terms of stakeholder intimacy and feedback and how the use of these media build reputation. Such a plan should show how competitors are using the same medium and describe the metrics for the success of the initiative.

Social Pages

The process of arriving at the right kind of content and interaction for a corporate social media page is not instantaneous but is the result of a partnership between a company and its stakeholders. The best analogue we can think of is the scientific concept called entrainment.

When fireflies gather at night, they gradually synchronize their flashing. When two pendulums are swinging in different rhythms, they eventually fall into the same rhythm. This is a process called "entrainment" and it is found throughout the natural world.

We think this is what is happening in social media or should be happening. Corporations signal what they are interested in and invite response. Fans indicate what they are interested in and invite response. The iteration of this process over time creates entrainment, in which both parties adjust to create discussions of mutual interest and benefit. In the case of pendulums, the faster one slows down and the slower one speeds up.

What this means is that companies need to start in social media with invitations of various kinds:

a. Ask stakeholders their opinion about a current topic
b. Post questions or comments with a link to your page on other sites and pages

    c.   Provide links to your Facebook page in as many places as possible—your own website, in advertising, off- and on-line

    d.   Make it appealing for your stakeholders to embed links to your fan page on their personal or institutional pages

As Marc Monseau, social media evangelist at Johnson & Johnson has said, if you treat social media as a broadcast channel, you are missing the point. The key to strengthening your brand through social media is to understand that your opportunity is to strengthen the brands of your stakeholders. Subconsciously and consciously, sometimes even with specific commercial intent, people use social media to build their own brands. When people post comments, links and photos, play games and make recommendations, they are signaling something about themselves—I'm a good mom, I listen to cool music, I know a lot about food, I'm an expert in urban planning, I care about Nicaragua, my favorite relaxation is watching Mad Men. The sum total of these signals represents an investment in a personal brand.

So whether a company makes bedroom slippers, iced coffee or life insurance, the issue is not how to market itself but figuring out how its products, services, insights and information can help consumers market their own personal brands. This means taking advantage of all the rich features that are enabled by social media such as widgets, polling, recommendation engines, games, "gifts" and other interactive tools.

Corporate Blogs

The best corporate blogs are often the expression of an individual's personality, sometimes that of the CEO, often that of one of its chief communicators. If a CEO is well known through traditional media and has a strong brand, it often makes sense for the corporate blog to be that of the CEO, but it is far more important that a corporate blog be effective than that it be by the CEO. Our research suggests that the best corporate blogs share the following characteristics:

- They are updated frequently, ideally no less than once a week
- They are usually focused on issues of interest to a company and its stakeholders but not confined to describing company activities
- They have an informal, "thinking out loud" character that plays against the conventional tone of corporate pronouncements
- They are generous in spirit, doling out praise and recognition, when appropriate, even to competitors
- However, they can be passionate in advocating the company's position when merited

We recognize that many corporations are concerned about negative commentary and do not allow comments on their corporate blogs. We believe this is a mistake and that only the most egregious inappropriate or ad hominem attacks should be suppressed. The benefits of receiving direct feedback outweigh the risks.

Dealing with Bloggers

Over the past five years, most industries and professions have seen the emergence of professional and semi-professional bloggers whose opinions are very influential. However, even though some of these bloggers comport themselves in a manner no different to traditional media, many have a more informal approach. This often extends to a sarcastic tone of voice, the use of anonymous sources and a distaste for embargoes. Bloggers often have little compunction in lambasting a communicator if they think the outreach has been ham-fisted. Every communication department will need to decide how best to handle these aspects of social media, but outreach to bloggers can be a very effective way to build reputation in a way that mirrors traditional media. Some specific differences need to be kept in mind:

a. As solo (usually) practitioners, bloggers are particularly sensitive to being scooped. If there are a small number of influential bloggers in your industry, be very careful about playing favorites

b. Unlike most conventional journalists, they measure success not by the prominence of their pieces but by other criteria, such as unique visitors, subscribers, their value to Google AdSense, their link strength and the number of comments their posts attract. Figuring out how to help them on these metrics can create positive relationships

Twitter

Next to LinkedIn, the social network for professionals, Twitter is the most popular social medium used by corporations, largely because its 140 character limit makes few demands on content creation. Too often, however, the Twitterfeed is seen as one more place to stuff with links to conventional press releases. The best corporate Twitterfeeds share the following characteristics:

*   They have an intra-day Tweet frequency no lower than 3–5
*   They manage different Twitterfeeds for different purposes—thought leadership, customer service, investor relations
*   They follow other Tweeters whose output is interesting to their stakeholders and re-Tweet content frequently
*   They ask questions

As with Facebook, the astute corporate Twitter user is treating the medium as a way to have a conversation not a broadcast. Jet Blue, Dell and FedEx are examples of companies using the medium effectively for customer service.

YouTube

Creating video content or a channel on YouTube is an effective way to deliver messages that showcase a new product, tackle a customer service issue or manage a crisis. Companies need to be prepared for users to mash up their video content with other material to mock them but this risk usually within manageable boundaries. Although the comment field in YouTube is frequently captive to the lowest common denominator, it still makes sense to comment on reasonable posts, even if they are negative. This will often bring our more supporters than detractors. In the YouTube environment, if a company is posting video of its executives, these videos need to be confident, approachable and informal. Since the corporate home page is often the first port of call for people looking for information, there should be links to YouTube content from that page, particularly on critical issues.

LinkedIn

LinkedIn is a valuable tool through which a company can stay in touch with its professional stake-holders—such as business partners, alumni and potential employees. The "Group" features are an increasingly effective way to obtain feedback to specific questions and share thought leadership. It is a particularly effective tool for business-to-business companies.

The field of social media is evolving constantly, with the current trend being for geo-targeting applications such as Four Square, through which members share location-based information. Innovations likely to endure are currently hard to distinguish from those that will wither away. Leadership in the corporate communication use of these platforms will go to those companies that experiment continuously to uncover the unique value of these emerging technologies.

Further Reading and Websites

Introduction

Part six provides an opportunity to acquire information, the raw material necessary for adaptation and change. The arc of business, as the economic downturn of 2008–2009 illustrated, goes down as well as up. Organizations are created and destroyed. The commercial landscape is littered with companies that grew, thrived, matured, withered, and died. Sustainable companies pass on the ability to thrive and survive. They nurture the capability of their professionals to see the future, to understand what it means, and to be unafraid to adapt and change to meet it.

Included in the list of books and articles are authors and ideas that have influenced and shaped the themes of this book. They demonstrate that corporate communication is both art and science. The combination creates a compelling and credible vision for what is ahead, informed by the past, but applied to the future. Individuals and companies that sustain themselves have the capacity to adapt, to change and meet new and unforeseen challenges. They have the ability to create a path where none existed. They have the ability to lead and "to see around the corner." And, if necessary, they can scrap current practices and start from scratch.

The list of websites offers a rich platform of focused sources of information and knowledge that serve to nourish professional development and growth. The categories include:

- Library Catalogs
- Professional Organizations and Societies
- Socially Responsible Investing
- Corporate Citizenship Information
- Communication Research Sites and Centers
- Investor Relations Information

- Intercultural Communication
- Media Sources
- Wire Services
- Database Sites
- Public Relations Firms and Agencies

Adaptation to the forces that are shaping contemporary business—globalization, Web 2.0 (3.0), and the networked enterprise—begins with a deep understanding of these forces and the knowledge of the strategies to harness them. Part 6 provides the tools and raw materials for companies and individuals to create sustainable corporate communication practices.

FURTHER READING

Adler, Gordon. "The Case of the Floundering Expatriate." *Harvard Business Review* (July/August 1995): 24–40.

Akerlof, George A., and Robert J. Shiller. *Animal Spirits: How Human Psychology Drives the Economy, and Why It Matters for Global Capitalism*. Princeton: Princeton University Press, 2009.

Alsop, Ron. *The 18 Immutable Laws of Corporate Reputation: Creating, Protecting, and Repairing Your Most Valuable Asset*. New York: Free Press, 2004.

———. *The Trophy Kids Grow Up*. New York: Wall Street Journal Books, 2008.

annenberg.usc.edu/AboutUs/News/090225SCPRCsurvey.aspx.

Argenti, Paul. *Corporate Communication*. 4th ed. New York: Irwin/McGraw-Hill, 2007.

The Authentic Enterprise: An Arthur W. Page Society Report. New York: Arthur W. Page Society, 2007. www.awpagesociety.com/images/uploads/2007AuthenticEnterprise.pdf.

Axtell, Roger. *Dos and Taboos Around the World*. 3rd ed. New York: Wiley, 1993.

Bakan, Joel. *The Corporation: The Pathological Pursuit of Profit and Power*. New York: Free Press, 2004.

Barbee, George, and Mark Lutchen. "Local Face, Global Body." *PW Review* (Spring 1995): 18–31.

Barnlund, Dean C. *Communicative Styles of Japan and America: Images and Realities*. Belmont, CA: Wadsworth, 1989.

Beatty, Jack. *Age of Betrayal: The Triumph of Money in America, 1865–1900*. New York: Alfred A. Knopf, 2007.

Belasen, Alan T. *The Theory and Practice of Corporate Communication: A Competing Values Perspective*. Los Angeles and London: Sage, 2008.

Benkler, Yochai. *The Wealth of Networks: How Social Production Transforms Markets and Freedom*. New Haven: Yale University Press, 2006.

Bennis, Warren, Daniel Goleman, and James O'Toole. *Transparency: How Leaders Create a Culture of Candor*. San Francisco, CA: Jossey-Bass, 2008.

Bhidé, Amar. *The Venturesome Economy*. Princeton: Princeton University Press, 2008.

Bolton, Roger. "The Audacity of Authenticity." www.awpagesociety.com/awp_blog/comments/the_audacity_of_authneticity, January 25, 2009.

Bronn, Peggy, and Roberta Berg, eds. *Corporate Communication: A Strategic Approach to Building Reputation*. 2nd ed. Oslo: Gyldendal, 2005.

Broom, Glen. *Cutlip and Center's Effective Public Relations*. Englewood Cliffs, NJ: Prentice-Hall, 2008.

Buffett, Warren. *A Plain English Handbook: How to Create Clear SEC Disclosure Documents*. Office of Investor Education and Assistance, U.S. Securities and Exchange Commission, Washington, DC.

CCI (Corporate Communication International) Corporate Communication Practices and Trends Studies 1999–2009. www.corporatecomm.org/studies.

Chappell, Tom. *Managing Upside Down: The Seven Intentions of Values-Centered Leadership*. New York: William Morrow, 1999.

———. *The Soul of a Business: Managing for Profit and the Common Good*. New York: Bantam, 1993.

Chernow, Ron. *The House of Morgan: An American Banking Dynasty and the Rise of Modern Finance*. New York: Simon & Schuster, 1991.

———. *Titan: The Life of John D. Rockefeller, Sr.* New York: Random House, 1998.

Cohan, William D. *The Last Tycoons: The Secret History of Lazard Frères & Co.* New York: Doubleday, 2007.

Cole, Benjamin Mark, ed. *The New Investor Relations: Expert Perspectives on the State of the Art*. New York: Bloomberg Press, 2004.

Collins, Jim. *Good to Great*. New York: HarperCollins, 2001.

———. *How the Mighty Fall, and Why Some Companies Never Give In*. New York: HarperCollins, 2009.

Connaughton, Stacey, and John Daly. "Long Distance Leadership: Communicative Strategies for Leading Virtual Teams." In David Pauleen, ed., *Virtual Teams: Projects, Protocols and Processes*, 116–143. Hershey, PA: Idea Group Publishing, 2004.

Connor, Daryl R. *Managing at the Speed of Change: How Resilient Managers Succeed and Prosper Where Others Fail*. New York: Villard Books, 1994.

Copeland, Lennie, and Lewis Griggs. *Going International: How to Make Friends and Deal Effectively in the Global Marketplace*. New York: Random House, 1985.

Cornelissen, Joep. *Corporate Communications: Theory and Practice*. London: Sage, 2004.

Culture Grams. Provo, UT: Center for International Studies, Brigham Young University. Set of 96 cultures.

Curtin, Patricia, and T. Kenn Gaither. *International Public Relations: Negotiating Culture, Identity, and Power*. Thousand Oaks, CA: Sage, 2007.

Cutlip, Scott M. *The Unseen Power: Public Relations, A History*. Hillsdale, NJ: Erlbaum Associates, 1994.

Cutlip and Center's Effective Public Relations. Englewood Cliffs, NJ: Prentice-Hall, 2008.

Deal, Terence E., and Allan A. Kennedy. *Corporate Cultures: The Rites and Rituals of Corporate Life*. 1982. New York: Perseus Books, 2000.

Denning, Stephen. *The Secret Language of Leadership: How Leaders Inspire Action Through Narrative*. San Francisco, CA: Jossey-Bass, 2007.

DiPiazza, Samuel S., Jr., and Robert G. Eccles. *Building Public Trust: The Future of Corporate Reporting*. New York: Wiley, 2002.

Directory of American Firms Operating in Foreign Countries. New York: World Trade Academy Press. (Annual.)

Directory of Foreign Firms Operating in the United States. New York: World Trade Academy Press. (Annual.)

Doorley, John, and Helio Fred Garcia. *Reputation Management: The Key to Successful Public Relations and Corporate Communication*. New York: Routledge, 2007.

Downes, John, and Gordon E. Goodman. *Dictionary of Finance and Investment Terms*. 7th ed. Hauppauge, NY: Barron's Educational Series, 2006.

Doyle, Edward. *How the United States Can Compete in the World Marketplace*. New York: IEEE, 1991.

Drucker, Peter. "The Coming of the New Organization." *Harvard Business Review* (January–February 1988): 45ff.

———. *Managing in a Time of Great Change*. New York: Truman Talley Books/Dutton, 1995.

The Dynamics of Public Trust in Business—Special Report. New York: The Arthur W. Page Society and the Business Roundtable Institute for Corporate Ethics, 2009.

Eccles, Robert G., et al. *The Value Reporting Revolution*. New York: Wiley, 2001.

Edwards, Mike. "A Broken Empire: After the Soviet Union's Collapse." *National Geographic* 183:3 (March 1993): 2–53.

Elliott, A. Larry. *How Companies Lie: Why Enron Is Just the Tip of the Iceberg*. New York: Crown Business, 2002.

Europe: World Partner—The External Relations of the European Community. Luxembourg: Office for Official Publications of the European Communities, 1991.

European Community in the Nineties. Washington, DC: EC Delegation to the United States, 1992.

European Community 1992 and Beyond. Luxembourg: Office for Official Publications of the European Communities, 1991.

Ewen, Stuart. *PR! A Social History of Spin*. New York: Basic Books, 1996.

Fearn-Banks, Kathleen. *Crisis Communications: A Casebook Approach*. 2nd ed. Mahwah, NJ: Lawrence Erlbaum Associates, 2002.

Ferraro, Gary. *The Cultural Dimension of International Business*. 2nd ed. Englewood Cliffs, NJ: Prentice-Hall, 1994.

Foster, Richard, and Sarah Kaplan. *Creative Destruction: Why Companies That Are Built to Last Underperform the Market—And How to Successfully Transform Them.* New York: Doubleday/Currency, 2001.

Four Minute Men Bulletin 1, May 22, 1917.

Frankel, Tamar. *Trust and Honesty: America's Business Culture at a Crossroad.* New York: Oxford University Press, 2006.

Frederick, Howard. *Global Communication and International Relations.* Belmont, CA: Wadsworth Publishing Company, 1993.

Frieden, Jeffrey A. *Global Capitalism: Its Fall and Rise in the Twentieth Century.* New York: W.W. Norton, 2006.

Friedman, George. *The Next 100 Years: A Forecast for the 21st Century.* New York: Doubleday, 2009.

Friedman, Milton. "The Social Responsibility of a Corporation Is to Increase Its Profits." *The New York Times Sunday Magazine,* September 13, 1970.

Friedman, Thomas L. *Hot, Flat, and Crowded: Why We Need a Green Revolution—And How It Can Renew America.* New York: Farrar, Straus, and Giroux, 2008.

———. *The Lexus and the Olive Tree: Understanding Globalization.* New York: Farrar, Straus, and Giroux, 2000.

———. "Why How Matters." *The New York Times,* October 15, 2008.

———. *The World Is Flat: A Brief History of the 21st Century.* New York: Farrar, Straus, and Giroux, 2005.

Fukuyama, Francis. *Trust: The Social Virtues and the Creation of Prosperity.* New York: Free Press, 1995.

Fussler, Claude, Aron Cramer, and Sebastian van der Vegt, eds. *Raising the Bar: Creating Value with the United Nations Global Compact.* Sheffield, UK: Greenleaf, 2004.

Galambos, Louis, and Joseph Pratt. *The Rise of the Corporate Commonwealth: U.S. Business and Public Policy in the Twentieth Century.* New York: Basic Books, 1988.

Gannon, Martin. *Understanding Global Cultures.* Thousand Oaks, CA: Sage, 1994.

Gates, Bill. *The Road Ahead.* Rev. ed. New York: Viking, 1996.

Gerstner, Louis V., Jr. *Who Says Elephants Can't Dance? Inside IBM's Historic Turnaround.* New York: HarperBusiness, 2002.

Gladwell, Malcolm. *Outliers: The Story of Success.* New York: Little, Brown, 2008.

———. *The Tipping Point: How Little Things Can Make a Big Difference.* New York: Little, Brown, 2002.

Goffee, Robert, and Gareth Jones. *The Character of a Corporation: How Your Company's Culture Can Make or Break Your Business.* New York: HarperBusiness, 1998.

———. "Why Should Anyone Be Led By You?" *Harvard Business Review* (September–October 2000).

Goode, Erica. "How Culture Molds Habits of Thought." *The New York Times,* August 8, 2000, D1, D2, D4.

Goodman, Michael B., ed. *Corporate Communication: Theory and Practice with Essays from the Conference on Corporate Communication.* Albany, NY: SUNY Press, 1994.

———. *Corporate Communication for Executives.* Albany, New York: SUNY Press, 1998.

Goodman, Michael B. "Corporate Communication Practice and Pedagogy at the Dawn of the New Millennium." *Corporate Communication: An International Journal* 11, no. 3 (2006): 196–213.

———. "The Special Section on Professional Communication in Russia: An American Perspective." IEEE *Transactions on Professional Communication* 37, no. 2 June. [?AU: what year?]

———. "Today's Corporate Communication Function." In *Handbook of Corporate Communication and Public Relations: Pure and Applied,* edited by Sandra Oliver, 200–26. London: Routledge, 2004.

———. *Work with Anyone Anywhere: A Guide to Global Business.* Belmont, CA: Professional Publications, 2006.

———. *Write to the Point: Effective Communication in the Workplace.* Englewood Cliffs, NJ: Prentice-Hall, 1984.

Gordon, John Steele. *The Business of America.* New York: Walker & Co., 2001.

Gottlieb, Marvin R., and Lori Conkling. *Managing the Workplace Survivors: Organizational Downsizing & the Commitment Gap.* Westport, CT: Quorum Books, 1995.

Gower, Karla. "US Corporate Public Relations in the Progressive Era." *Journal of Communication Management* 12, no. 4 (2008): 305–18.

"Graduate Education." In *The Professional Bond: Report of the Commission on Public Relations Education,* 51–52. November 2006. www.commpred.org.

Grunig, James E., Larissa A. Grunig, and David Dozier. *Excellent Public Relations and Effective Organizations: A Study of Communication Management in Three Countries.* Lea's Communication Series, 2002.

"Guidelines for Measuring Relationships in Public Relations." www.instituteforpr.com.

"Guidelines for Restoring Public Trust in Corporations." www.corporatecomm.org/pdf/PRCoalitionPaper_9_11Final.pdf.

Hackos, JoAnn, and Dawn M. Stevens. *Standards for Online Communication: Publishing Information for the Internet/World Wide Web/Help Systems/Corporate Intranets.* New York: Wiley, 1997.

Haglund, E. "Japan: Cultural Considerations." *International Journal of Intercultural Relations* 8 (1984): 61–76.

Hall, Edward T. *Beyond Culture.* New York: Doubleday, 1976.

———. *The Dance of Life.* Garden City, NY: Anchor/Doubleday, 1987.

———. *Hidden Differences: Doing Business with Japan.* Garden City, NY: Anchor/ Doubleday, 1987.

———. *The Hidden Dimension.* Garden City, NY: Doubleday, 1966.

———. *The Silent Language.* Garden City, New York: Doubleday, 1959.

Hall, Lynne. *Latecomer's Guide to the New Europe: Doing Business in Central Europe.* New York: American Management Association, 1992.

Hamel, G., and C.K. Pralahad. "The Core Competence of the Corporation." *Harvard Business Review* (1990): 79–91.

Hammer, Michael, and James Champy. *Reengineering the Corporation.* New York: HarperBusiness, 2006.

Handy, Charles. *The Hungry Spirit: Beyond Capitalism—A Quest for Purpose in the Modern World.* London: Arrow Books, 1998.

———. *Understanding Organizations.* 4th ed. London: Penguin, 1993.

Harrington, Matt. "2009 Edelman Trust Barometer." CCI Symposium on Reputation—Trust Me?" March 2009. www.corporatecomm.org/archive.

Harrison, E. Bruce. *Corporate Greening 2.0: Create and Communicate Your Company's Climate Change & Sustainability Strategies.* Exeter, NH: Publishing Works, 2008.

Heath, Chip, and San Heath. *Made to Stick: Why Some Ideas Survive and Some* Die. New York: Random House, 2007.

Henzier, Herbert. "The New Era of Eurocapitalism." *Harvard Business Review* (July–August 1992): 57–68.

Hiebert, Ray. *Courtier to the Crowd: The Story of Ivy Lee and the Development of Public Relations.* Ames: Iowa State University Press, 1966.

Hodgson, Kent. "Adapting Ethical Decisions to a Global Marketplace." *Management Review* (May 1992): 53–57.

Hofstede, Geert. *Cultures and Organizations.* London: HarperCollins, 1991.

———, and Gert Jan Hofstede. *Cultures and Organizations: The Software of the Mind.* 2nd ed. New York: McGraw-Hill, 2005.

Hollander, Edwin P. *Inclusive Leadership.* New York: Routledge, 2009.

Hurst, David K. *Crisis & Renewal: Meeting the Challenge of Organizational Change.* Boston: Harvard Business School Press, 1995.

Hutton, James, Michael B. Goodman, Jill Alexander, and Christina Genest. "Reputation Management: The New Face of Corporate Public Relations?" *Public Relations Review* 27 (2001): 247–61.

Ihlen, Oyvind, Betteke van Ruler, and Magnus Frederiksson, eds. *Public Relations and Social Theory: Key Figures and Concepts.* New York: Routledge, 2009.

Jarvis, Jeff. *What Would Google Do?* New York: Collins Business, 2009.

Jeffries-Fox, Bruce. "Advertising Value Equivalency." www.instituteforpr.com.

Keller, Ed, and Jon Berry. *The Influentials.* New York: Free Press, 2003.

Kelly, Alan. *The Elements of Influence: The New Essential System of Managing Competition, Reputation, Brand, and Buzz.* New York: Dutton, 2006.

"Key Competences for Lifelong Learning: European Reference Framework." Luxembourg: Office for Official Publication of the European Communities, 2007.

Kidder, Tracy. *The Soul of a New Machine.* New York: Avon Books, 1981.

Kotter, John P. *The Heart of Change: Real-Life Stories of How People Change Their Organizations.* Boston: HBS Press, 2002.

Kouzes, James, and Barry Posner. *The Leadership Challenge.* 3rd ed. San Francisco, CA: Jossey-Bass, 2002.

KPMG Ethics and Compliance Report. 2008.

Krugman, Paul. *The Return of Depression Economics and the Crisis of 2008.* New York: Norton, 2009.

Laskin, Alexander V. "A Descriptive Account of the Investor Relations Profession." *Journal of Business Communication* 46 (April 2009): 208–33.

Lawrence, Paul, and Charalambos Vlachoutsicos. "Joint Ventures in Russia: Put the Locals in Charge." *Harvard Business Review* (January–February 1993): 44–54.

Lebow, Rob, and William Simon. *Lasting Change: The Shared Values Process.* New York: Van Nostrand Reinhold, 1997.

Lessig, Lawrence. *Remix: Making Art and Commerce Thrive in the Hybrid Economy.* New York: Penguin, 2008.

Levitt, Arthur. *Take on the Street.* New York: Pantheon Books, 2002.

Lewis, Michael. *Liar's Poker: Rising Through the Wreckage on Wall Street.* New York: Penguin, 1990.

Lind, Michael. *Up from Conservatism: Why the Right Is Wrong for America.* New York: Simon & Schuster, 1996.

Little, Jeffrey B. *Understanding Wall Street.* 4th ed. New York: McGraw-Hill, 2004.

"The Long Demise of Glass-Steagall: A Chronology Tracing the Life of the Glass-Steagall Act from Its Passage in 1933 to Its Death Throes in the 1990s." www.pbs.org/wgbh/pages/frontline/shows/wallstreet/weill/demise.html.

Low, Jonathan, and Pan Cohen Kalafut. *Invisible Advantage: How Intangibles Are Driving Business Performance.* Cambridge, MA: Perseus Publishing, 2002.

Mamet, David. *Glengarry Glen Ross.* New York: Grove, 1983.

Martin, Dick. *Rebuilding Brand America: What We Must Do to Restore Our Reputation and Safeguard the Future of American Buisiness.* New York: AMACOM, 2007.

———. *Secrets of the Marketing Masters.* New York: AMACOM, 2009.

———. *Tough Calls: AT&T and the Hard Lessons Learned from the Telecom Wars.* New York: AMACOM, 2005.

McRae, Hamish. *The World in 2020: Power, Culture and Prosperity.* Boston: Harvard Business School Press, 1994.

Micklethwait, John, and Adrian Wooldridge. *The Company: A Short History of a Revolutionary Idea.* New York: The Modern Library, 2003.

Miller, Arthur. *Death of a Salesman.* New York: Penguin, 1949.

The Mission in the Marketplace: How Responsible Investing Can Strengthen the Fiduciary Oversight of Foundation Endowments and Enhance Philanthropic Missions. Social Investment Forum Foundation, Washington, DC, 2009. www.socialinvest.org.

Moore, Geoffrey. *Crossing the Chasm: Marketing and Selling Technology Products to Mainstream Customers.* New York: Harper, 1991.

Morgan, Garth. *Images of Organization.* Rev. ed. Thousand Oaks, CA: Sage, 2006.

Morley, Michael. *How to Manage Your Global Reputation: A Guide to the Dynamics of International Public Relations.* New York: NYU Press, 2002.

Newman, Paul, and A.E. Hotchner. *Shameless Exploitation: In Pursuit of the Common Good.* New York: Doubleday, 2003.

Noal, Emile. *Working Together—The Institutions of the European Community.* Luxembourg: Office for Official Publications of the European Communities, 1994.

Nolan, Richard L., and David Croson. *Creative Destruction: A Six-Stage Process for Transforming the Organization.* Boston: Harvard Business School Press, 1995.

Ott, J. Steven. *The Organizational Culture Perspective.* Pacific Grove, CA: Brooks/Cole Publishing, 1989.

Ottman, Jacquelyn A. *Green Marketing: Opportunity for Innovation.* 2nd ed. Chicago: NTC Business Books, 1997.

Pagell, Ruth, and Michael Halperin. *International Business Information: How to Find It, How to Use It.* Phoenix, AZ: ORYX, 1994.

Parsons, Patricia. *Ethics in Public Relations: A Guide to Best Practice.* London: Kogan Page, 2005.

Penning, Timothy. "First Impressions: US Media Portrayals of Public Relations in the 1920s," *Journal of Communication Management* 12, no. 4 (2008):. 358.

Peters, Glen. *Waltzing with the Raptors: A Practical Roadmap to Protecting Your Company's Reputation.* New York: Wiley, 1999.

Pocket Pal. 19th ed. International Paper, 2003.

Private Sector Summit on Public Diplomacy: Models for Action. PR Coalition White Paper 2007. www.corporatecomm.org.

Putnam, Robert. *Bowling Alone: The Collapse and Revival of American Community.* New York: Simon & Schuster, 2000.

Quirke, Bill. *Communicating Corporate Change: A Practical Guide to Communicating Corporate Strategy*. New York: McGraw-Hill, 1996.

Redefining Work 2. London: RSA, 2003.

"Restoring Trust in a Cynical American Public—Francis Fukuyama." In *Arthur W. Page Society Journal, 2002 Annual Conference: Earning Trust: Aligning Communications and Leadership Behavior*, edited by Edwin Nieder. New York: Arthur W. Page Society, 2002.

Ries, Al, and Laura Ries. *The Fall of Advertising & the Rise of PR*. New York: HarperBusiness, 2002.

Rowland, D. *Japanese Business Etiquette: A Practical Guide to Success in the Global Market Place*. New York: Praeger, 1986.

Russell, Karen, and Carl Bishop. "Understanding Ivy Lee's Declaration of Principles." Presented at Association for Education in Journalism and Mass Communication, Chicago, August 2008. www.allacademic.com/meta/p272004_index.html.

Schmertz, Herb, and William Novak. *Goodbye to the Low Profile: The Art of Creative Confrontation*. Boston: Little, Brown, 1986.

Schumpeter, Joseph A. *Capitalism, Socialism and Democracy*. New York: Harper & Row, 1942.

Scott, David L. *How Wall Street Works*. 2nd ed. New York: McGraw-Hill, 1999.

Seidman, Dov. *How: Why HOW We Do Anything Means Everything . . . in Business (and in Life)*. New York: Wiley, 2007.

Semler, Richardo. "Who Needs Bosses?" *Across the Board* (February 1994).

Shirky, Clay. *Here Comes Everybody: The Power of Organizing without Organizations*. New York: Penguin Press, 2008.

Soros, George. *The New Paradigm for Financial Markets: The Credit Crisis of 2008 and What It Means*. New York: Public Affairs, 2008.

———. *On Globalization*. New York: Public Affairs, 2002.

Stine, Harry. *The Corporate Survivors*. New York: AMACOM, 1986.

Surowiecki, James. *The Wisdom of Crowds: Why the Many Are Smarter Than the Few and How Collective Wisdom Shapes Business, Economies, Societies, and Nations*. New York: Doubleday, 2004.

Tapscott, Don. *Growing Up Digital*. New York: McGraw-Hill, 1998.

———. *Grown Up Digital: How the Net Generation Is Changing the World*. New York: McGraw-Hill, 2009.

———, and Anthony D. Williams. *Wikinomics: How Mass Collaboration Changes Everything*. New York: Penguin Group, 2006.

Terpstra, V., and K. David. *The Cultural Environment of International Business*. Cincinnati: South Western, 1995.

Tomasco, Robert M. *Rethinking the Corporation: The Architecture of Change*. New York: AMACOM, 1993.

Tomorrow's Company: The Role of Business in a Changing World. London: RSA (Royal Society for the Encouragement of Arts, Manufactures & Commerce), 1994.

Tompkins, Phillip K. *Apollo, Challenger, Columbia: The Decline of the Space Program; A Study in Organizational Communication*. Los Angeles: Roxbury Pub. Co., 1993.

———. *Organizational Communication Imperatives: Lessons of the Space Program*. Los Angeles: Roxbury Pub. Co., 1993.

Toward Greater Transparency: Modernizing the Securities and Exchange Commission's Disclosure System—21st Century Disclosure Initiative: Staff Report. Washington, DC: SEC, 2009.

Trompenaars, Fons. *Riding the Waves of Culture: Understanding Cultural Diversity in Business*. London: The Economist Books, 1993.

United Nations Global Compact. www.unglobalcompact.org/Portal/Default.asp.

U.S. Securities and Exchange Commission. "Interactive Data to Improve Financial Reporting." 17 CFR Parts 229, 230, 232, 239, 240, and 249.

van Riel, Cees B.M. "Defining Corporate Communication." In *Corporate Communication*, edited by Peggy Simcic Brown and Roberta Wiig Berg. Oslo: Gyldendal, 2005.

van Ruler, Betteke, Ana Vercic, and Dejan Vercic, eds. *Public Relations Metrics: Research and Evaluation*. New York: Routledge, 2008.

Victor, David. *International Business Communication*. New York: HarperCollins, 1992.

Watson, Tom. *Causewired: Plugging in, Getting Involved, Changing the World*. Hoboken, NJ: Wiley, 2009.

Weiss, Stephen. "Negotiating with 'Rornans.'" *Sloan Management Review* (Winter 1994): 51–61.

Wheatley, Margaret J. *Leadership and the New Science: Discovering Order in a Chaotic World*. San Francisco, CA:

Berrett-Koehler, 1999.

Wyld, David C. *The Blogging Revolution: Government in the Age of Web 2.0.* Washington, DC: IBM Center for the Business of Government, 2007.

Zakaria, Fareed. *The Post-American World.* New York: Norton, 2008.

Zittrain, Jonathan. *The Future of the Internet and How to Stop It.* New Haven: Yale University Press, 2008.

WEBSITES

LIBRARY CATALOGS

The Library of Congress Online Catalog. Access to the holdings of the National Library.
http://catalog.loc.gov
The New York Public Library. Online search of library catalogs and, with a library card, free databases.
http://catnyp.nypl.org/

PROFESSIONAL ORGANIZATIONS AND SOCIETIES

Advertising Women of New York (AWNY). Founded in 1912 as first women's association in the communications industry. It supports career advancement of female practitioners and fosters the use of public relations to benefit the goals of business and society.
http://www.awny.org/
American Management Association. A membership organization devoted to management issues including communication.
http://www.amanet.org
The American Society of Corporate Secretaries. Incorporated under New York State's Not-for-Profit Corporation Law on November 6, 1946, this is a professional association whose membership is composed principally of corporate secretaries, assistant secretaries, and other persons who are involved in duties traditionally associated with the corporate secretarial function. Members are involved in such matters as corporate governance, records management, the regulation and trading of securities, proxy solicitation and other shareholder activities, and the administration of the office of the corporate secretary.
http://www.ascs.org
The Arthur W. Page Society. A professional organization with a single mission: to strengthen the management policy role of the chief corporate relations officer.
http://www.awpagesociety.com
Association for Business Communication (ABC) . An international organization committed to fostering excellence in business communication scholarship, research, education, and practice.
http://www.businesscommunication.org/
Business Ethics. The website for *Business Ethics* magazine. Its mission: to promote ethical business practices, to serve the growing community of professionals striving to live and work in responsible ways, and to help create financially healthy companies in the process.
http://www.business-ethics.com
Business for Social Responsibility (BSR). The mission of this membership organization is to help its member companies achieve long-term commercial success by implementing policies and practices that honor high ethical standards and meet their responsibilities to all who are impacted by their decisions.
http://www.bsr.org
The Business Roundtable. An association of chief executive officers of leading U.S. corporations with a combined workforce of more than 10 million in the United States. The Roundtable is committed to advocating public policies that foster vigorous economic growth, a dynamic global economy, and a well-trained and productive U.S. workforce essential to future competitiveness.
http://www.businessroundtable.org/

Center for Media & Democracy. A nonprofit, public interest organization dedicated to investigative reporting on the public relations industry.
http://www.prwatch.org

The Chartered Institute of Public Relations (CIPR). CIPR was founded in 1948 and currently comprises over 8,000 members. Formerly the Institute of Public Relations, it was granted a Royal Charter in February 2005. Its primary goals include leading the public relations profession and providing the highest quality of service, advocating high ethical standards and promoting the profession. It is the largest organization of its kind in Europe.
http://www.cipr.co.uk/

Communications Roundtable. The association of 24 public relations, marketing, graphics, advertising, training, information technology, and other communications organizations with more than 12,000 professional members.
http://www.roundtable.org

The Conference Board. An international business membership organization whose mission is to improve the business enterprise system and to enhance the contribution of business to society.
http://www.conference-board.org

The Conference Board Council on Communications Strategy. The Council is a forum for off-the-record discussion focused on key communications issues and state-of-the-art management practice. Through the exchange of ideas and knowledge, the group seeks to enhance the professional development of its members and improve management of the corporate communication function. Members also advise the board on its communications research and meeting program.
http://www.conference-board.org/councils/councilsDetailUS.cfm?Council_ID=50

Corporate Communication International (CCI) at Baruch College/CUNY—Devoted to the theory and practice of corporate communication. CCI provides vital information for corporate practitioners as well as scholars, policy makers, and the general public.
http://www.corporatecomm.org/

Council of Communication Management (CCM). The Council of Communications Management was established more than 40 years ago as a forum for seasoned professionals to share best practices in organizational communications. CCM's network of several hundred senior communicators, representing companies and consultancies of all sizes, confronts strategic communications issues every day.
http://www.ccmconnection.com/

Council of Public Relations Firms (CPRF). Formed in 1998 by dozens of America's leading public relations firms. Their goal was to create the first national association to represent the interests of public relations firms. Many of those firms remain members today.
http://www.prfirms.org

The European Public Relations Confederation (CERP). CERP was founded in 1959 by practitioners in Belgium, France, Germany, Italy, and The Netherlands. It includes all major national public relations associations in Europe representing a total of about 22,000 public relations practitioners, consultants, in-house specialists, teachers, researchers, and students. The main objective of CERP is to establish and maintain contact among its associations and members.
http://www.cerp.org/

The European Public Relations Education and Research Association (EUPRERA). The main goal of EUPRERA is to stimulate knowledge and the practice of public relations education and research in Europe, with the exchange and communication of knowledge among its members as paramount.
http://www.euprera.org/

FEI Financial Executives International. The mission of FEI is to be the preeminent association for financial executives, to alert members to emerging issues, to develop the professional and management skills of members, to provide forums for peer networking, to advocate the views of financial executives, and to promote ethical conduct. FEI is the professional association of choice for senior-level corporate financial executives and the leading organization dedicated to advancing ethical, responsible financial management. Representing 15,000 individuals, FEI has been the voice of corporate finance for over 70 years.
http://www.financialexecutives.org/

Global Alliance. The Alliance enhances networking opportunities for professionals and serves as a vehicle for examining ethical standards, universal accreditation options, and other initiatives to strengthen the influence of the public relations industry among its constituents around the world.

http://www.globalpr.org/

IABC (International Association of Business Communicators). A global membership organization offering programs and products for people and organizations in the fields of public relations, employee communication, marketing communication, and public affairs.

http://www.iabc.com

IABC Research Foundation. IABC funds worldwide research that supports and advances the communication profession by delivering knowledge, findings, and tools that are vital to successful business communication. The foundation has funded research that balances both practitioner information needs and emerging and future concerns of the profession.

http://www.iabc.com/fdtnweb/index.html

IEEE Professional Communication Society. Fosters a community dedicated to understanding and promoting effective communication in engineering, scientific, and other technical environments. Its mission is to advance technical and scientific communication as an essential element of engineering; promote and disseminate best practices and research results on the development, maintenance, delivery, and management of technical content; and promote and facilitate leading-edge education and training of engineers, scientists, and other technically oriented professionals in communication theory and practice.

http://www.ieeepcs.org/

Institute for Public Relations (IPR). Established originally as the Foundation for Public Relations Research and Education, promotes and encourages academic and professional excellence.

http://www.instituteforpr.com

Institute for Public Relations Measurement Commission. The mission of the Measurement Commission is to be the leading provider of information about and advocate for PR and related communication research and evaluation.

http://www.instituteforpr.org/about/measurement_commission/

IPR. A London-based organization dedicated to high standards of professionalism and continuous professional development. See CIPR.

http://www.ipr.org.uk

IPRA. International Public Relations Association, London.

http://www.ipranet.org

The Issue Management Council (IMC). The professional membership organization for people whose work is managing issues and those who wish to advance the discipline.

http://www.issuemanagement.org/

Marcom Exchange. An interactive community developed by and for marketing communication professionals.

http://www.marcomexchange.com

The National Association of Corporate Directors (NACD). An educational, publishing, and consulting organization in board leadership and the only membership association for boards, directors, director-candidates, and board advisors.

http://www.nacdonline.org

National Black Public Relations Society. The National Black Public Relations Society was established to benefit top PR and affiliated services professionals. It aims to address the needs of the global society and to prepare future PR professionals. Its mission is to address the challenges and emphasize the opportunities for the diversified constituency it serves through education, expansion, and empowerment.

http://www.nbprs.org/

National School Public Relations Association (NSPRA). NSPRA's mission is to advance education through responsible communication.

http://www.nspra.org/

Prime Point Foundation. A nonprofit registered public trust founded in December 1999 in Madras, India, with these goals: to create public relations and communication awareness among management students, corporate professionals in both the public and private sectors, politicians, etc.; and to enhance the professional skills for public relations practitioners, journalists, and other communicators.

http://www.primepointfoundation.org

PRSA Counselors Academy. Counselors Academy provides the public relations industry's premier one-to-one personal and professional development opportunities, mentoring, and inspiration to the most senior-level practitioners in PR firms.

http://www.counselorsacademy.org/

PRSA Foundation. Founded in 1990, the PRSA Foundation is the philanthropic arm of the public relations profession and particularly PRSA. Its charge is threefold: demonstrate the knowledge and practice of public relations; further understanding among business leaders about the power and value of public relations; and identify and develop future professionals, enhancing the diversity of the profession.

http://www.prsafoundation.org/

Public Affairs Council (PAC). Professional organization for public affairs executives.

http://www.pac.org

Public Relations Society of America (PRSA). PRSA provides a forum for addressing issues affecting the profession and the resources for promoting the highest professional standards.

http://www.prsa.org/

RSA, The Royal Society for the Encouragement of Arts, Manufactures & Commerce. Founded in 1754 to embolden enterprise, to enlarge science, to refine art, to improve manufactures and to extend commerce. An independent, nonaligned, multidisciplinary registered charity with more than 20,000 Fellows from all walks of life.

http://www.rsa.org.uk

SOCIALLY RESPONSIBLE INVESTING *(See also Investor Relations and Corporate Citizenship)*

Center for Responsible Business. Launched in 2003, the Center for Responsible Business's vision is to create a more sustainable, ethical, and socially responsible society by establishing the Haas School of Business as the preeminent educational institution for research, teaching, experiential learning, and community outreach in areas of Corporate Social Responsibility (CSR).

http://www.haas.berkeley.edu/responsiblebusiness/

The Center for Social Philanthropy. Aims to provide an online portal of research, resources, and tools for foundations and donors seeking to maximize the long-term, social, and environmental impact of their philanthropic work, not only through grant-making but also by leveraging the full range of assets at their disposal.

http://www.socialphilanthropy.org/

Center on Corporate Responsibility (ICCR). ICCR is a leader of the corporate social responsibility movement. ICCR's membership is an association of 275 faith-based institutional investors, including national denominations, religious communities, pension funds, foundations, hospital corporations, economic development funds, asset management companies, colleges, and unions. ICCR and its members press companies to be socially and environmentally responsible. Each year ICCR-member religious institutional investors sponsor over 200 shareholder resolutions on major social and environmental issues.

http://www.iccr.org/

Coalition of Community Development Financial Institutions. Formed in 1992 as an ad hoc policy development and advocacy initiative, the Coalition of Community Development Financial Institutions (CDFI Coalition) is the lead national organization in the United States promoting the work of community development financial institutions (CDFIs).

http://cdfi.org/

Community Investing Center. The Center's mission is to provide financial professionals with information and resources to help them channel more money into community investing. This includes "how-to" guidance for investors and the most comprehensive database of Community Investment Institutions (CIIs).

http://communityinvest.org/

Corporate Social Responsibility Initiative (Harvard Business School). Grounded in Harvard Business School's mission to educate leaders who make a difference in the world, the Social Enterprise Initiative aims to inspire, edu-

cate, and support current and emerging leaders in all sectors to apply management skills to create social value.
http://www.hbs.edu/socialenterprise/

Foundation Partnership on Corporate Responsibility (FPCR). Created in 1996, the purpose of FPCR is to facilitate and provide technical assistance to foundations that want to be more active as shareholders on social and environmental issues.
http://www.foundationpartnership.org/

Research Initiative on Social Entrepreneurship (RISE). A research project at Columbia Business School whose mission is to study and disseminate knowledge about the markets, metrics, and management of for-profit and nonprofit social enterprise and social venturing.
http://www.riseproject.org/

Social Funds. SocialFunds.com features over 10,000 pages of information on Socially Responsible Investing (SRI) mutual funds, community investments, corporate research, shareowner actions, and daily social investment news.
http://www.socialfunds.com/

Social Investment Forum (SIF). The only national membership association dedicated to advancing the concept, practice, and growth of socially and environmentally responsible investing (SRI). Members integrate economic, environmental, social, and governance factors into their investment decisions, and SIF provides programs and resources to advance this work.
http://socialinvest.org/

Sustainable Endowments Institute. A nonprofit organization engaged in research and education to advance sustainability in campus operations and endowment practices. Founded in 2005, the institute is a special project of Rockefeller Philanthropy Advisors.
http://www.endowmentinstitute.org/

The Tellus Institute. Formed in 1976 as a not-for-profit research and policy organization, Tellus is an international leader in assessing critical environment and development issues. The Institute has conducted thousands of projects throughout North America and the rest of the world.
http://tellus.org/

United Nations Environment Programme Finance Initiative (UNEP FI). A unique global partnership between the United Nations Environment Program (UNEP) and the private financial sector. UNEP FI works closely with over 160 financial institutions who are signatories to the UNEP FI Statements and a range of partner organizations to develop and promote linkages between the environment, sustainability, and financial performance. Through regional activities, a comprehensive work program, training programs, and research, UNEP FI carries out its mission to identify, promote, and realize the adoption of best environmental and sustainability practice at all levels of financial institution   operations.
http://www.unepfi.org/

United Nations Principles for Responsible Investment. PRI provides a framework for environmental, social, and corporate governance (ESG) issues that can affect the performance of investment portfolios.
http://www.unpri.org/

CORPORATE CITIZENSHIP INFORMATION

AccountAbility. A nonprofit organization established in 1995 to promote accountability innovations that advance responsible business practices and the broader accountability of civil society and public organizations. Its 350 members include businesses, NGOs, and research bodies, and elect an international council that includes representatives from Brazil, India, North America, Russia, South Africa, and Europe. AccountAbility has created the AA1000 Sustainability Assurance and Stakeholder Engagement Standards, the Responsible Competitiveness Index covering the links between responsible business practices and the competitiveness of over 80 countries, and, in collaboration with CSRNetwork, the Accountability of the world's largest companies published annually with *Fortune International*.
http://www.accountability21.net/

Aspen Institute. The mission of the Aspen Institute is to foster enlightened leadership, the appreciation of timeless ideas and values, and open-minded dialogue on contemporary issues. Through seminars, policy programs, conferences, and leadership development initiatives, the institute and its international partners seek to promote the pursuit of com-

mon ground and deeper understanding in a nonpartisan and non-ideological setting.
http://www.aspeninstitute.org/

Business Civic Leadership Center (BCLC). The Business Civic Leadership Center (BCLC) is a 501(c)3 affiliate of the U.S. Chamber of Commerce, the world's largest business federation. BCLC is the U.S. Chamber's resource and voice for businesses and their social and philanthropic interests. BCLC was founded in May 2000 as the Center for Corporate Citizenship (CCC). It stemmed from the U.S. Chamber of Commerce's informal creation of corporate public service coalitions and the fact that corporate citizenship was an emerging area of business management strategy.
http://www.uschamber.com/bclc/default

Center for Corporate Citizenship at Boston College. Provides leadership in establishing corporate citizenship as a business essential, so all companies act as economic and social assets to the communities they impact by integrating social interests with other core business objectives. Through its research, executive education, consultation, and convenings on issues of corporate citizenship, the center is the leading organization helping corporations define their role in the community. Part of the Carroll School of Management, the 16-year-old center has nearly 350 member companies, a full-time staff of 30, and has trained over 5,000 executives in its various courses.
http://www.bc.edu/centers/ccc/index.html

Citizen Works. A nonprofit, nonpartisan, 501(c)(3) organization founded by Ralph Nader in April 2001 to advance justice by strengthening citizen participation in power.
http://www.citizenworks.org

CSR Europe. A business-to-business network for Corporate Social Responsibility in Europe. Its mission is to help companies achieve profitability, sustainable growth, and human progress by placing Corporate Social Responsibility in the mainstream of business practice.
http://www.csreurope.org

CSRWire Corporate Social Responsibility Newswire. CSRWire seeks to promote the growth of corporate responsibility and sustainability through solutions-based information and positive examples of corporate practices. Its core services are distribution of press releases, links to corporate reports, promotion of CSR events, and access to CSR resources.
http://www.CSRwire.com/

Ethical Corporation. An independent business publication for corporate responsibility, producing 12 issues per year and dedicated to providing companies around the world with practical advice and examples of how to successfully integrate responsible corporate practice into management systems.
http://www.ethicalcorp.com

Ethics Resource Center (ERC). A nonprofit, nonpartisan educational organization whose vision is an ethical world. The mission of the Ethics Resource Center is to be a leader and a catalyst in fostering ethical practices by individuals and institutions.
http://www.ethics.org

The Foundation Center. Seeks to support and improve institutional philanthropy by promoting public understanding of the field and helping grantseekers succeed. It collects, organizes, and communicates information on U.S. philanthropy; conducts and facilitates research on trends in the field; and provides education and training on the grant-seeking process. The center is the nation's leading authority on institutional philanthropy and is dedicated to serving grantseekers, grantmakers, researchers, policymakers, the media, and the general public.
http://www.fdncenter.org

The Global Reporting Initiative (GRI). Established in late 1997 with the mission of developing globally applicable guidelines for reporting on the economic, environmental, and social performance, initially for corporations and eventually for any business, governmental, or non-governmental organization (NGO). Convened by the Coalition for Environmentally Responsible Economies (CERES) in partnership with the United Nations Environment Programme (UNEP), the GRI incorporates the active participation of corporations, NGOs, accountancy organizations, business associations, and other stakeholders from around the world.
http://www.globalreporting.org

Institute for Global Ethics. Founded to promote ethical behavior in individuals, institutions, and nations through research, public discourse, and practical action.
http://www.globalethics.org/

The Prince of Wales International Business Leaders Forum. An international educational charity set up in 1990 to promote responsible business practices that benefit business and society and that help to achieve social, economic, and environmentally sustainable development, particularly in new and emerging market economies.
http://www.iblf.org/

Ron Brown Award for Corporate Leadership. The only presidential award to honor companies for the exemplary quality of their relationships with employees and communities. This annual award is presented to companies that have demonstrated a deep commitment to innovative initiatives that not only empower employees and communities but also advance strategic business interests.
http://www.ron-brown-award.org/

Social Accountability International (SAI). A charitable human rights organization dedicated to improving workplaces and communities by developing and implementing socially responsible standards. The first standard to be fully operational is Social Accountability 8000 (SA8000), a workplace standard that covers all key labor rights and certifies compliance through independent, accredited auditors.
http://www.sa-intl.org

The SPIN (Strategic Press Information Network) Project. Provides comprehensive media training, intensive media strategizing, and resources to community organizations across the country. SPIN helps grow the capacity of grassroots groups to shape public opinion and garner positive media attention. The project believes that there is a direct correlation between a community's improved media skills and its ability to get good press.
http://www.spinproject.org/

Transparency International. An international non-governmental organization devoted to combating corruption, Transparency International brings civil society, business, and governments together in a global coalition. TI raises awareness about the damaging effects of corruption, advocates policy reform, works toward the implementation of multilateral conventions, and subsequently monitors compliance by governments, corporations, and banks. TI does not expose individual cases; it focuses on prevention and reforming systems. A principal tool in the fight against corruption is access to information.
http://www.transparency.org

The World Business Council for Sustainable Development (WBCSD). A coalition of 160 international companies united by a shared commitment to sustainable development through the three pillars of economic growth, ecological balance, and social progress.
http://www.wbcsd.org

COMMUNICATION RESEARCH SITES AND CENTERS

The American Enterprise Institute for Public Policy Research. A private, nonpartisan, not-for-profit institution dedicated to research and education on issues of government, politics, economics, and social welfare. AEI's purposes are to defend the principles and improve the institutions of American freedom and democratic capitalism—limited government, private enterprise, individual liberty and responsibility, vigilant and effective defense and foreign policies, political accountability, and open debate.
http://www.aei.org

The Brookings Institution. An independent, nonpartisan organization devoted to research, analysis, education, and publication focused on public policy issues in the areas of economics, foreign policy, and governance. The goal of Brookings activities is to improve the performance of American institutions and the quality of public policy by using social science to analyze emerging issues and to offer practical approaches to those issues in language aimed at the general public.
http://www.brookings.edu

The Cato Institute. A nonprofit public policy research foundation. It seeks to broaden the parameters of public policy debate to allow consideration of the traditional American principles of limited government, individual liberty, free markets, and peace. Toward that goal, the institute strives to achieve greater involvement of the intelligent, concerned lay public in questions of policy and the proper role of government.
http://www.cato.org

Center for Communication. Produces on- and off-site programs in every field of media: TV, radio, newspapers, publishing, film, public relations, advertising, and digital technologies.

http://www.cencom.org

The Center for Public Integrity. A nonprofit, nonpartisan, tax-exempt organization that conducts investigative research and reporting on public policy issues in the United States and around the world. The center was founded in 1989 by Charles Lewis following a career in network television news. Through thorough, thoughtful, and objective analyses, the center hopes to serve as an honest broker of information and to inspire a better-informed citizenry to demand a higher level of accountability from its government and elected leaders.

http://www.publicintegrity.org/

The Fanning Center for Business Communication. Established in the autumn of 1990. In 1998, the center, its faculty, and its programs became a part of the Department of Management in the Mendoza College of Business. The center hosts a conference on corporate communication each fall.

http://www.nd.edu/~fanning/

The Independent Media Institute (IMI). A nonprofit organization dedicated to strengthening and supporting independent and alternative journalism and to improving the public's access to independent information sources. IMI believes democracy is enhanced and public debate broadened as more voices are heard and points of view made available. IMI has four editorial and service components: AlterNet.org, a public interest online magazine; AlterNet Syndication, a news service for the independent press; WireTap, an online magazine for socially conscious youth; and the Strategic Press Information Network (SPIN), which trains grassroots and advocacy groups in communication skills. More information about these entities can be found below.

http://www.independentmedia.org/

The Museum of Public Relations. Established in 1997 as a place to go to learn about how ideas are developed for industry, education, and government, and how they have been applied to successful public relations programs since the PR industry was born.

http://www.prmuseum.com

The Pew Research Center. An independent opinion research group that studies attitudes toward the press, politics, and public policy issues, best known for regular national surveys that measure public attentiveness to major news stories, and for polling that charts trends in values and fundamental political and social attitudes. The center's purpose is to serve as a forum for ideas on the media and public policy through public opinion research. In this role it serves as an important information resource for political leaders, journalists, scholars, and public interest organizations. All current survey results are made available free of charge.

http://people-press.org/

Silver Anvil Resource Center. Developed as an online resource for professionals to search for best practices in public relations from the PRSA Silver Anvil archive.

http://www.silveranvil.org

INVESTOR RELATIONS INFORMATION

American Stock Exchange. By continuously cultivating new ideas and building new relationships across the globe, the American Stock Exchange creates financial opportunities for both individual and institutional investors and for issuers spanning every industry sector and market size. This site offers investors and issuers access to market and historical data, charts and tools, and news and education available from the only primary exchange to offer trading in three distinct lines of business: a wide variety of listed equities; an extensive options market; and an unrivaled listing of more than 100 exchange-traded funds (ETFs), the securities category pioneered by the American Stock Exchange.

http://www.amex.com/

Association for Enterprise Opportunity (AEO). A national leadership organization and the voice of microenterprise development. By providing cutting-edge training, knowledge sharing, federal and state public policy and advocacy, and communications, AEO empowers a community of nearly 500 member organizations to be uniquely effective in serving the needs of microentrepreneurs who do not have access to traditional sources of business education or capital.

http://www.microenterpriseworks.org/

Ceres. A national network of investors, environmental organizations, and other public interest groups working with companies and investors to address sustainability challenges such as global climate change.
http://www.ceres.org/

Clean Energy Group (CEG). A nonprofit, 501(c)(3) organization dedicated to greater use of clean energy technologies in the United States and abroad through innovation in finance, technology, and policy. CEG operates as a "market assist" nonprofit catalyst to improve clean energy markets. CEG works with public fund managers, private investors, business academics, and other energy consultants to provide information, advocacy, and analysis to develop market opportunities for clean energy.
http://www.cleanegroup.org/

The Corporate Library. The Corporate Library is intended to serve as a central repository for research, study, and critical thinking about the nature of the modern global corporation, with a special focus on corporate governance and the relationship between company management, their boards, and shareholders. Most general content on the site is open to visitors at no cost; advanced research relating to specific companies and certain other advanced features are restricted to subscribers only.
http://www.thecorporatelibrary.com/

The Council of Institutional Investors. An organization of large public, labor funds and corporate pension funds that seeks to address investment issues that affect the size or security of plan assets. Its objectives are to encourage member funds, as major shareholders, to take an active role in protecting plan assets and to help members increase return on their investments as part of their fiduciary obligations.
http://www.cii.org

Domini 400 Social Index (DSI). Established in 1990 as the benchmark for measuring the impact of social screening on financial performance.
http://www.kld.com/indexes/ds400index/index.html

Domini Social Investment LLC. The Domini is an investment firm specializing exclusively in socially responsible investing. The firm manages funds for individual and institutional investors who wish to integrate social and environmental standards into their investment decisions.
http://www.domini.com/

Edgar Online. The website for information on Securities and Exchange Commission filings.
http://www.edgar-online.com

Financial Accounting Standards Board. The mission is to establish and improve standards of financial accounting and reporting for the guidance and education of the public, including issuers, auditors, and users of financial information.
http://www.fasb.org

House Committee on Financial Services. The committee oversees the entire financial services industry, including the securities, insurance, banking, and housing industries. The committee also oversees the work of the Federal Reserve, the Treasury, the SEC, and other financial services regulators.
http://financialservices.house.gov/

Institute for Responsible Investment (IRI). IRI convenings, research, and activities promote and expand the field of responsible investment. The IRI works with investors, corporations, public sector organizations, and research institutes to coordinate thinking and action around issues of strategic importance to long-term wealth creation for shareholders and society.
http://www.bccc.net/responsibleinvestment

The Investor Network on Climate Risk (INCR). A network of institutional investors and financial institutions that promotes better understanding of the financial risks and investment opportunities posed by climate change.
http://www.incr.com/

IRRC (Investor Responsibility Research Center). The world's leading source of impartial, independent research on corporate governance, proxy voting, and corporate responsibility issues. IRRC's mission is to provide the highest-quality research on companies and shareholders worldwide.
http://www.bapd.org/ginler-1.html

KLD Research & Analytics, Inc. An independent investment research firm providing management tools to professionals integrating environmental, social, and governance factors (ESG) into their investment decisions.

http://www.kld.com/

NASDAQ. Diffuse, decentralized, and open stock market model committed to open electronic architecture. A network of networks consisting of broker-dealers, traders, electronic communications networks, and various order-routing systems.

http://www.nasdaq.com

National Investor Relations Institute (NIRI). A professional association of corporate officers and investor relations consultants responsible for communication among corporate management, the investing public, and the financial community.

http://www.niri.org

The National Community Investment Fund. A nonprofit, private equity trust that invests in banks, thrifts, and credit unions that generate both financial and social returns. These Community Development Banking Institutions (CDBIs) (a term used by NCIF to describe depository institutions with a community development focus) may be located in urban, rural, or Native American markets, and may be minority owned, minority focused, or majority owned. However, to be considered a CDBI, an institution must focus a substantial part of its business on low- to moderate-income people or communities.

http://www.ncif.org/

NYSE Euronext (The New York Stock Exchange). Mission: to add value to the capital-raising and asset-management process by providing the highest-quality and most cost-effective, self-regulated marketplace for the trading of financial instruments, promote confidence in and understanding of that process, and serve as a forum for discussion of relevant national and international policy issues.

http://www.nyse.com

Proxy Democracy. Provides tools to help investors use their voting power to produce positive changes in the companies they own. It is a nonprofit, nonpartisan project supported by foundations that are themselves interested in being responsible investors.

http://proxydemocracy.org/

Sid Cato's Annual Report Website. Monitors the world's annual reports to shareholders of publicly held companies, using proprietary computer programs for independent and consistent appraisal.

http://www.sidcato.com/

Stock Market Yellow Pages. Offers a list of exclusively public companies with a particular word or phrase in their description. Many sites link to further research and provide either a "symbol lookup" that lists symbols of companies, or a description search that provides an incomplete or cluttered list of both private *and* public companies (e.g., Business.com). Stock Market Yellow Pages indexes a "buzz word" (e.g., "biomedical") of a hot industry to get a list comprised solely of companies for research.

http://www.StockMarketYellowPages.com/

The U.S. Securities and Exchange Commission (SEC). The primary mission of the SEC is to protect investors and maintain the integrity of the securities markets. As more and more first-time investors turn to the markets to help secure their futures, pay for homes, and send children to college, these goals are more compelling than ever. The laws and rules that govern the securities industry in the United States derive from a simple and straightforward concept: all investors, whether large institutions or private individuals, should have access to certain basic facts about an investment prior to buying it. To achieve this, the SEC requires public companies to disclose meaningful financial and other information to the public.

http://www.sec.gov

U.S. Senate Committee on Banking, Housing and Urban Affairs. " . . . to which committee shall be referred all proposed legislation, messages, petitions, memorials and other matters relating to the following subjects: banks, banking, and financial institutions; control of prices of commodities, rents and services; deposit insurance; economic stabilization and defense production; export and foreign trade promotion; export controls; federal monetary policy, including the Federal Reserve System; financial aid to commerce and industry; issuance and redemption of notes; money and credit, including currency and coinage; nursing home construction; public and private housing (including veterans housing); renegotiation of government contracts; urban development and urban mass transit."

http://www.senate.gov/~banking

INTERCULTURAL COMMUNICATION

Intercultural Communication Institute (ICI). A private, nonprofit foundation designed to foster an awareness and appreciation of cultural differences in both the international and domestic arenas. ICI is based on the beliefs that (1) education and training in the areas of intercultural communication can improve competence in dealing with cultural difference and thereby minimize destructive conflict among national, ethnic, and other cultural groups; and (2) we therefore share an ethical commitment to further education in this area.
http://www.intercultural.org

Intercultural Press. Publisher of books exploring and celebrating cultural diversity and the experiences of working and studying abroad. Here can be found invaluable resources to help in developing skills in intercultural communication and cultivating fulfilling personal and professional relationships abroad.
http://www.interculturalpress.com

SIETAR. The Society for Intercultural Education, Training and Research is the world's largest interdisciplinary network for professionals and students working in the field of intercultural relations. The primary purpose of SIETAR is to encourage the development and application of values, knowledge, and skills that promote and reinforce beneficial and long-lasting intercultural relations at the individual, group, organization, and community levels.
http://www.sietar.org

MEDIA SOURCES

ABC
http://www.abc.com

AlterNet.org. A project of the Independent Media Institute, a nonprofit organization dedicated to strengthening and supporting independent and alternative journalism. First launched in 1998, AlterNet's online magazine provides a mix of news, opinion, and investigative journalism on subjects ranging from the environment, the drug war, technology, and cultural trends, to policy debate, sexual politics, and health issues. AlterNet features 12 Special Coverage Areas, each with its own hub site, editor, and weekly newsletter.
http://www.alternet.org

AP Associated Press
www.ap.org

BBC
http://news.bbc.co.uk/

Bloomberg
http://www.bloomberg.com

BusinessWeek
http://www.businessweek.com

CBS
http://www.cbs.com

CNN
http://www.cnn.com

CNNmoney
http://money.cnn.com/

The Center for Public Integrity. A nonprofit organization dedicated to producing original, responsible investigative journalism on issues of public concern. The center is nonpartisan and nonadvocacy, committed to transparent and comprehensive reporting both in the United States and around the world.
http://www.publicintegrity.org/

The Economist
http://www.economist.com

The Financial Times
http://www.ft.com

Forbes

http://www.forbes.com

Fortune

http://www.fortune.com

The Holmes Report. Provides insight and intelligence to public relations professionals through research, industry White Papers, and a weekly newsletter.

http://www.holmesreport.com/

MediaChannel.org. A nonprofit, public interest website dedicated to global media issues. MediaChannel offers news, reports, and commentary from an international network of media-issues organizations and publications, as well as original features from contributors and staff. Resources include thematic special reports, action toolkits, an indexed directory of hundreds of affiliated groups, and a search engine constituting the single largest online media-issues database. MediaChannel is concerned with the political, cultural, and social impacts of the media, large and small. MediaChannel exists to provide information and diverse perspectives and inspire debate, collaboration, action, and citizen engagement.

http://www.mediachannel.org

National Public Radio (NPR)

http://www.npr.org

NBC

http://www.nbc.com

The New York Times

http://www.nytimes.com

O'Dwyer's. Publisher of industry-related directories and newsletters.

http://www.odwyerpr.com

Poynter.org. A site that exists to help journalists do their jobs better and serve their communities. Advances the goals of the Poynter Institute by making its expertise, teaching, and research accessible worldwide.

http://www.poynter.org

PR News

http://www.prandmarketing.com

PR Watch. A quarterly publication of the Center for Media & Democracy, it is dedicated to investigative reporting on the public relations industry. It serves citizens, journalists, and researchers seeking to recognize and combat manipulative and misleading PR practices. It specializes in blowing the lid off today's multi-billion dollar propaganda-for-hire industry, naming names, and revealing how public relations wizards concoct and spin the news, organize phony "grassroots" front groups, spy on citizens, and conspire with lobbyists and politicians to thwart democracy.

http://www.prwatch.org/

PR Week

http://www.prweekus.com

ProPublica. An independent, nonprofit newsroom that produces investigative journalism in the public interest.

http://www.propublica.org/

The Ragan Report. U.S. publication for the PR industry.

http://www.ragan.com

Schuster Institute for Investigative Journalism. The nation's first investigative reporting center, based at Brandeis University, was launched in September 2004 to help fill the void in high-quality public interest and investigative journalism—and to counter the increasing corporate control of what Americans read, see, and hear. The goal is to investigate significant social and political problems and uncover corporate and government abuses of power.

http://www.brandeis.edu/investigate/

The Wall Street Journal

http://www.wsj.com

United States Newspapers. Directory of United States Newspapers, with links to over 3,300 newspapers .

http://www.50states.com/news/

WIRE SERVICES

Business Wire
http://www.businesswire.com
PRNewswire
http://www.prnewswire.com
Reuters
http://www.reuters.com

DATABASE SITES

About.com. A network of websites on hundreds of topics.
http://www.about.com
Corporate Information. Lists of other sites with information on private and international companies.
http://www.corporateinformation.com
Hoover's Online. The website of the international publisher of business information and company profiles.
http://www.hoovers.com
Lexis Nexis. A database service with more than a billion documents from over 8,000 databases.
http://www.lexisnexis.com
Thomas Register. Information and detailed descriptions of products and services. Searches company listings, brand names, and catalogs.
http://www.thomasregister.com

PUBLIC RELATIONS FIRMS AND AGENCIES

Lists of Public Relations Firms and Agencies are available online at many sites, including:
Council of Public Relations Firms
http://www.prfirms.org
Public Relations Society of America
http://www.prsa.org

APPENDIX

A Corporate Communication Workshop

PART I

A Strategy Summit

Introduction

Successful corporate communication, as we have discussed, is based on a strong combination of key elements: a clear strategic vision for the organization's future, consensus among key decision makers, a communication program aligned to business priorities, and an agreed-upon company vocabulary for expressing all of these ideas. This portfolio is simple to describe but often unnervingly difficult to execute over the long term. This difficulty can stem from a number of sources. Among the most pernicious, we would certainly count:

- Failure to agree upon a strategic corporate direction
- Inadequate measures for determining progress in this direction
- Poorly understood or conflicting uses of language to describe both process and goals
- Ambiguity in the business/communication roles for individual leaders of the organization
- Disagreement about priorities and, therefore, obstacles to effective execution
- Failure to align the communication program with business operations
- Inadequate communication outward from the leadership team about the strategic direction

Communication leaders and management teams have met these challenges in a variety of ways. In our experience, one of the most effective ways of doing this is through a facilitated corporate communication workshop or strategy summit that brings together corporate leadership, the communication team, and other relevant players to lay the foundations for a communication program that is effective and sustainable.

Strategy Summit

Depending on the complexity of the corporate image issues to be addressed, the strategy summit can be accomplished in as little as half a day or be extended over two days. Ideally, corporate management and the communications team can set aside at least one full day for the discussion without interruptions from cell phones or PDAs. This is one reason why strategy summits are often best conducted off-site.

It is also quite important to appoint a facilitator from within the company or to retain an outside facilitator to orchestrate the discussion. Each of these two choices has advantages and disadvantages. The insider will be very familiar with the company and with the personalities in the room, since this can help prevent blind allies in the discussion or interpersonal conflict. On the other hand, the insider may not feel comfortable standing up to senior management where necessary, and be unwilling to ask the "stupid" question. The outsider may be hampered by lack of intimate knowledge of the company and the personalities of its executives but, by contrast, is less likely to be intimated or distracted by interpersonal history or baggage. The one skill needed by the facilitator is, naturally, to be able to command the room. Given two equally gifted candidates, our recommendation would be to select the outsider. Too much knowledge can be an obstacle to honest debate and fresh thinking.

Workshop Flow

Expectations

Unless all the participants are well known to each other, the meeting should start with introductions. Even when those assembled have worked closely together for many years, it is nonetheless very important to follow introductions by asking everyone present to describe briefly what they hope to get out of the workshop, both in terms of their participation and, quite specifically, what outcomes or outputs they are expecting. It is not important that everyone's expectations for content and outcomes be the same. What is important is that these hopes and expectations be voiced publicly at the outset of the meeting. If there is a broad spectrum of different outcomes, this enables the facilitator to begin by getting the team assembled to prioritize the hoped-for outcomes and gain consensus from the whole group. It is also a good way to begin to identify divergent terminologies and ensure that every member of the group is using words that mean the same thing.

Passion

In most instances, the first exercise in identifying expectations for the meeting also begins to expose the passions that individual members of the group experience about the organization. The second exercise is designed to propel this thinking toward the next level by asking every participant to contribute his or her thoughts on three key questions:

- What makes you passionate about the organization?
- What do you think the organization currently does supremely well?
- What do you think the organization currently does not do very well?

The individual contributions to these questions engage the participants in an emotionally and psychologically charged way. What quickly emerges in most instances, however, is a cross-section of opinions about what business the organization is really in. Identifying and developing consensus about what business a company is truly in is a crucial first step in putting together a communication strategy. The job of the facilitator is not only to capture all of these inputs but also to move the discussion in the direction of the critical question that needs to be answered in order to move the communication strategy. It is common for participants in strategy summits to begin their contributions by describing the features of their company's products or services. A manufacturer, for example, might describe his company as being in the business of providing headphones and dwelling on the key feature: their noise-canceling technology. Prompted by the facilitator, he might acknowledge that the attributes of the product are enhanced sound quality across a spectrum of signal strengths. What the facilitator is steering him toward, in the classic marketing model, is an understanding of the ultimate benefit that his customer derives from use of the product. This can vary by customer type, but to complete the example, we could describe the benefit to a call center customer as being enhanced customer satisfaction as a result of more accurate order-taking, made possible because the sales reps can really hear what call-in customers are saying. By aggregating the benefits as seen by the participants, the facilitator begins to outline the key messages that will go into defining the corporate brand. In the case of a conglomerate with many businesses, the facilitator uses this stage of the summit to identify strengths held in common by all of the business lines.

Gap Analysis

By their nature, the inputs from the "passion" phase tend toward the positive side, which is why it is helpful to follow it with a gap analysis. Through a gap analysis, as its name suggests, the group, led by the facilitator, candidly assesses the extent to which the company's strengths are either a reality or might be described more charitably as aspirational. Having established what insiders honestly feel about the company's strengths and weaknesses, the gap analysis then proceeds to assess the same measures for external stakeholders. If this information is not known to the participants, a decision to finance a research project to assess this could be an outcome of the meeting. Otherwise, this information can be supplied by participants or sourced by the facilitator from another part of the organization in advance.

At this stage of the strategy summit, the facilitator will have created an increasingly rich and realistic picture of the company from an internal and external perspective. In order to proceed to a discussion of a corporate communication strategy, the next stage of the summit turns outward to a review of the market opportunity.

Market Opportunity

The preceding discussion is focused predominantly on the status quo, but commercial markets are in perpetual flux. To borrow an image from ice hockey legend Wayne Gretzky, companies need to figure out not just where the puck is but how to skate to where it's going. This analysis is designed to elicit from the participants their views about the benefits that customers and other stakeholders are looking for not just in the immediate term but for the foreseeable future. Depending on the industry segment in question, these future needs could be either relatively obvious or quite obscure. The important part of this exercise is not to be too limited in imagining possible scenarios. In most companies, the result of this exercise will be to highlight areas in which the organization is well positioned to supply the market's future needs, as well as others, where significant investments—either in operations or in marketing/sales—may need to be made to align it with the opportunity. The communication strategy to be developed will depend significantly on the ratio between these poles.

Message Platform

By this stage of the process, the team should have established a clear picture of the capabilities of the organization in the context of the future market opportunity as well as how those capabilities are seen by stakeholders. At this point, the facilitator can begin to point the participants toward creating a message platform that best encapsulates the current and future strengths of the company. The facilitator needs to steer this discussion of a message platform in order to identify messages that are

- Credible
- Relevant
- Differentiated

This is often the most challenging part of the workshop for two reasons. The first lies in the nature of the training of most business executives, who tend to have at least a passing acquaintance with the marketing discipline. As a result, when asked for possible corporate messages, they often default to taglines or advertising slogans, when what is called for is an underlying message rather than ad copy. The second is that in most industries it is very difficult to differentiate the benefits of one company or product set from another, and creating a message platform that is unique to one company is a significant achievement.

Telling the Story

Once a credible, relevant, and differentiated message platform has been created, the facilitator will press the group for stories, examples, and even the careers of individual employees that are apt illustrations of the message platform. Some aspect of innovation is usually a part of the message platform. In the case of a global oil company, part of the message platform involved the innovation created through the passionate commitment of employees. To illustrate this principle, the company unearthed an engineer who, in his own time, had developed a uniquely formulated engine lubricant suitable for use on the space shuttle. The strategy summit itself is usually only able to scratch the surface of the story-sourcing exercise, which then becomes the responsibility of the communication team.

Aligning the Enterprise

Once the message platform has been chosen and some of the potential narratives that demonstrate how the company "lives" the message have been identified, the facilitator leads a discussion to develop a plan to align the corporate brand with the platform. How comprehensive this plan becomes will depend on the size of the company and the gap between the company's existing message platform and the outgrowth from the summit. In most organizations, this plan will embrace changes to the website and corporate materials, new employee orientation content, educational materials for employees, a new take on the investor presentation, and so on. At the very least, the communication team will audit relations with all of the company's stakeholders—employees, customers, investors, regulators, and communities—to determine what needs to be changed. Sometimes the exercise will trigger an actual name change and even occasionally a new ticker symbol. In other situations, the summit will make it clear that the company needs to abandon certain communication practices and initiate others, all as part of aligning its corporate identity with the new message platform. The vast majority of this work takes place outside the confines of the summit itself, but it is crucial to have initial consensus with senior management in order to embark in this new direction.

Measurement and Accountability

The final phase of the summit is more than a wrapping up of loose ends. Indeed, it is often the process of choosing measurable objectives for the change initiative and assigning individual responsibilities that reveal whether the participants are truly convinced of the new direction and willing to stand behind it. A skilled facilitator will quickly recognize a summit that becomes derailed at this stage and can usually salvage some key components and persuade the group to meet again to iron out the remaining issues. Even if the summit has been extraordinarily successful, it is still crucial to set a calendar of follow-up meetings and, shortly after the summit, to publish a clear timeline of execution steps.

Conclusion

A well-run strategy summit can be a powerful tool for creating consensus for major change or to unearth the true communication drivers of a company's success. It is a useful tool to consider whether the organization has reached a real identity crisis or is simply trying to refresh a successful communication strategy. It is equally effective for companies perched on entering a new market or for a major market leader trying to evolve long-standing customer relationships. The key constituents to a successful summit are a strong facilitator, a culture of candor, and the imagination to play with the unknown.

PART II

Tactical Planning

The following exercise, we have found, is very useful in an orientation session for new staff. It provides an opportunity to experience the range of responsibility and activity for corporate communication.

This is a group exercise designed to emphasize the leadership role of corporate communicators. At the risk of oversimplifying the wealth of comment, data, and opinion on leadership, we can define a leader as "someone you choose to follow to a place you would not go by yourself."

Three very brief descriptions of companies follow. The companies are quite different, yet each shares the need to communicate with its various internal and external audiences.

Each group is assigned a company. Your group is asked to develop a communication strategy and plan for the company assigned to it. You will have time to meet with members of your group to discuss the issues and how you might address them in your communication strategy and implement them in your plan. Think about these issues in your plan:

- Issues and Common Practices in Corporate Communication
- Leadership and Corporate Communication
- The Value of Reputation

- PR Firms and Corporate Clients
- Media Relations and Corporate Communication
- The Corporation in Crisis
- Building a Corporate Communication Career

- Leadership in Corporate Communication Management
- Corporate Communication and Technologies
- Strategic Uses of Research: Measuring Corporate Reputation

- Ethics and Corporate Communication Practice
- Corporate Communication and the Changing Media Environment
- Investor Relations and the Capital Markets
- Building Trust in a Global Marketplace
- Gaining and Keeping a Seat at the Table: Counseling the CEO

At the end of the session your group will present the results of your thinking on the issues and how they apply, or do not apply, to your company. Each group will have 20 minutes for the presentation.

GROUP A

J.J. APPLESEED
(edited from the company website)

In 1901, John Johnson Appleseed (or "J.J.") founded his business on a love of nature, a profound belief in honesty, and a commitment to quality and customer satisfaction. These values remain the key to our company's longevity.

The Stores

With a lively history spanning nearly a century, J.J. Appleseed now has annual sales of more than $1.5 billion. A one-man direct mail order business has expanded to a 4,500 person operation with retail stores in Maine, Maryland, Massachusetts, New Jersey, Pennsylvania, and Virginia; mall and factory outlet stores on the East Coast; a website that draws visitors from around the world; a worldwide catalog business; and more than 20 retail stores in Japan and Europe.

The Service

Customer service is at the heart of J.J. Appleseed. The J.J. Appleseed Flagship Store remains on the site of J.J.'s original store, built circa 1908. To improve service, J.J. removed the locks on the store's front door more than 50 years ago and threw away the keys. Since then, customers have come and gone at all hours of the day and night, every day of the year.

Our Customer Satisfaction Department was created in the same spirit and fields calls around the clock. Our state-of-the-art Order Fulfillment Center can process up to 27 million items a year, ensuring prompt delivery to our customers.

The Great Outdoors

J.J. Appleseed offers more than just great products. To help customers get the most out of their outdoor adventures, our website features tips on fitness, hiking, biking, and kayaking. Our Wilderness & Outdoor Schools offer classes on everything from cross-country skiing to fly fishing. And to simplify your trip planning, J.J. Appleseed's website (Park Search) has the latest on activities at more than 2,200 state and national parks.

Some Numbers

Last year . . .

- 4,000 year-round employees
- Over 21,000 different items available (1,000 field testers)
- 14.5 million customer contacts; 179,000 on busiest day
- 16 million packages shipped
- 3 million visitors to our flagship store
- Over 16,000 customers participated in more than 800 Wilderness & Outdoor School classes
- Peak holiday season:
  —Over 11,000 seasonal employees
  —4,000 customer service representatives
  —85,000 online orders on busiest day
  —218,000 packages shipped on busiest day

Please note that J.J. Appleseed is a privately held, family-owned company. We release limited financial and operational information and do not produce an annual report.

Our Values

Much has changed since J.J. was born in a small town in Maine in 1875. Our founder was raised to believe in simple values. Nature was something to be revered. Family was a priority. Being neighborly was a necessity, and "do unto others" was not just a saying, but also a way of life. J.J. launched his company with one single product: an all-weather shoe. He believed strongly in fairness and value and made this saying the foundation of his business: "Sell good merchandise at a reasonable profit, treat your customers like human beings, and they will always come back for more."

J.J.'s values live on, and at J.J. Appleseed today, we still measure success by the satisfaction of our customers, which is why we've never wavered from the guarantee upon which our company was founded. We want to bring you the best products at the fairest price, backed by the best service possible—on the web, through our catalogs, and in our stores. In our experience, that's the only way to create relationships that stand the test of time. The Chairman of the Board recently stated, "A lot of people have fancy things to say about Customer Service, but it's just a day-in, day-out, ongoing, never-ending, persevering, compassionate kind of activity."

GROUP B

SPRUCEGOOSE CORPORATION (SGC)
(edited from the company website)

Sprucegoose Corporation (SGC) is an advanced technology company formed in 1995 by the merger of two of the world's premier technology companies.

Headquartered in Hidden Hills, California, Sprucegoose employs about 140,000 people worldwide and is principally engaged in the research, design, development, manufacture, and integration of advanced technology systems, products and services. Sprucegoose is led by William Boatman, Chairman and Chief Executive Officer, and Steven Robinson, President and Chief Operating Officer.

Customer Base

As a lead systems integrator and information technology company, the majority of Sprucegoose's business is with the U.S. Department of Defense and the U.S. federal government agencies. In fact, Sprucegoose is the largest provider of IT services, systems integration, and training to the U.S. Government. The remaining portion of Sprucegoose's business is comprised of international government and some commercial sales of our products, services, and platforms.

Financial Performance

The corporation reported 2006 sales of $39.6 billion, a backlog of $75.9 billion, and free cash flow of $3.8 billion.

Organization

Sprucegoose's operating units are organized into broad business areas.

- Aeronautics, with approximately $11.4 billion in 2006 sales, includes tactical aircraft, airlift, and aeronautical research and development lines of business.

- Space Systems, with approximately $7.9 billion in 2006 sales, includes space launch, commercial satellites, government satellites, and strategic missiles lines of business.

- Systems & IT Group, with approximately $20.3 billion in 2006 sales, will leverage our existing and emerging capabilities to address customers' growing needs for highly integrated systems and solutions. This includes missiles and fire control, naval systems, platform integration, federal services, energy programs, government and commercial IT, and aeronautical/aerospace services lines of business.

The corporation's New York Stock Exchange symbol is SPG, and its web address is www.sprucegoose.com.

Our Vision

Powered By Innovation, Guided By Integrity, We Help Our Customers Achieve Their Most Challenging Goals.

Our Values

Do What's Right
Respect Others
Perform with Excellence

Business Areas

Aeronautics
Space Systems
Systems & IT Group

2006 Sales: $39.6 Billion

Backlog: $75.9 billion

Employees: 140,000 employees in the United States and internationally

Operations: 939 facilities in 457 cities and 45 states throughout the United States; Internationally, business locations in 56 nations and territories

Headquarters: Sprucegoose Corporation (SGC)
Hidden Hills, CA USA

(888) 555-5050

GROUP C

RX WORLDWIDE INC. (RXW)
(edited from the company website)

RX Worldwide Inc. is dedicated to better health and greater access to healthcare for people and their valued animals. Our purpose is to help people live longer, healthier, and happier lives. Our route to that purpose is through discovering and developing breakthrough medicines; providing information on prevention, wellness, and treatment; consistent high-quality manufacturing of medicines and consumer products; and global leadership in corporate responsibility. Every day we help 38 million patients, employ more than 87,000 colleagues, utilize the skills of more than 12,000 medical researchers, and work in partnership with governments, individuals, and other payers for healthcare to treat and prevent illnesses—adding both years to life, and life to years.

Our Mission

We will become the world's most valued company to patients, customers, colleagues, investors, business partners, and the communities where we work and live.

Our Purpose

We dedicate ourselves to humanity's quest for longer, healthier, happier lives through innovation in pharmaceutical, consumer, and animal health products.

Our Values

To achieve our Purpose and Mission, we affirm our values of Integrity, Respect for People, Customer Focus, Community, Innovation, Teamwork, Performance, Leadership, and Quality.

Integrity. We demand of ourselves and others the highest ethical standards, and our products and processes will be of the highest quality.

Respect for People. We recognize that people are the cornerstone of RX Worldwide's success. We value our diversity as a source of strength, and we are proud of RX Worldwide's history of treating people with respect and dignity.

Customer Focus. We are deeply committed to meeting the needs of our customers, and we constantly focus on customer satisfaction.

Community. We play an active role in making every country and community in which we operate a better place to live and work, knowing that the ongoing vitality of our host nations and local communities has a direct impact on the long-term health of our business.

Innovation. Innovation is the key to improving health and sustaining RX Worldwide's growth and profitability.

Teamwork. We know that to be a successful company we must work together, frequently transcending organizational and geographical boundaries to meet the changing needs of our customers.

Performance. We strive for continuous improvement in our performance, measuring results carefully, and ensuring that integrity and respect for people are never compromised.

Leadership. We believe that leaders empower those around them by sharing knowledge and rewarding outstanding individual effort. Leaders are those who step forward to achieve difficult goals, envisioning what needs to happen and motivating others.

Quality. Since 1849, the RX Worldwide name has been synonymous with the trust and reliability inherent in the word Quality. Quality is ingrained in the work of our colleagues and all our Values. We are dedicated to the delivery of quality healthcare around the world. Our business practices and processes are designed to achieve quality results that exceed the expectations of patients, customers, colleagues, investors, business partners, and regulators. We have a relentless passion for Quality in everything we do.

OUR COMPANY

CEO:	J.D. Smith
RX Worldwide Locations:	**Corporate Headquarters:**
	New York, NY (USA)
	Research & Development:
	New Haven, Connecticut
	Warwick, England
	Nagano, Japan
	Paris, France
	San Jose, California
	Boston, Massachusetts
	Detroit, Michigan
	Consumer Health Care:
	Morristown, New Jersey
Web Site Address:	www.RXWorldwide.com
Stock Exchange Listings:	New York Stock Exchange (RXW)
	London (RXZ)
	Euronext
Number of Employees Worldwide:	87,000
2006 Revenues:	$48.371 billion
2006 Actual R&D Spending:	$7.599 billion
Key RX Worldwide Pharmaceutical Products:	Links to list of pharmaceutical products
Key RX Worldwide Consumer Health Care Products:	Links to list of consumer health care products.
Key RX Worldwide Animal Health Products:	Links to list of animal health products

INDEX

About the Authors

MICHAEL B. GOODMAN is Professor at Baruch College/City University of New York, where he is Director of the M.A. Program in Corporate Communication. He is Founder and Director of CCI Corporate Communication International. He is also Adjunct Professor of Corporate Communication at Fairleigh Dickinson University, and Visiting Professor of Corporate Communication at Aarhus School of Business (Denmark), University of Johannesburg (South Africa), Bangkok University, and Hong Kong Polytechnic University. He has published widely, including *Work with Anyone Anywhere: A Guide to Global Business* and *Corporate Communication for Executives*. He is on the Editorial Advisory Board and is Associate Editor for North America of *Corporate Communication: An International Journal*. He has been a consultant to more than 40 corporations and institutions on corporate communication, managerial communication, problem-solving, new business proposals, change, and corporate culture.

PETER B. HIRSCH has more than 25 years' experience in counseling global corporations and runs an independent consulting firm specializing in corporate reputation and issues management. He has also worked with foreign governments, including the governments of Greece, Colombia, and the Philippines. Previously, he was a partner at Porter Novelli, where he established the corporate communication practice and served as Global Practice Leader for Corporate Affairs. He has been Adjunct Professor teaching courses on a range of corporate communication topics at Baruch College/City University of New York, Columbia University, and Fordham University. He has also lectured at Fairleigh Dickinson University and New York University. He has written numerous articles, including "The Ulysses Project" (*Journal of Business Strategy*) and "My Country Is Different" (*Corporate Communication, An International Journal*). He is a member of the advisory board of Corporate Communication International and a member of the Public Relations Society of America.